Vital Minimum

Vital Minimum

NEED, SCIENCE, AND POLITICS IN MODERN FRANCE

Dana Simmons

The University of Chicago Press CHICAGO & LONDON

DANA SIMMONS is associate professor of history
at the University of California, Riverside.

The University of Chicago Press, Chicago 60637
The University of Chicago Press, Ltd., London

Printed in the United States of America

24 23 22 21 20 19 18 17 16 15 1 2 3 4 5

ISBN-13: 978-0-226-25156-1 (cloth)
ISBN-13: 978-0-226-25173-8 (e-book)
DOI: 10.7208/chicago/9780226251738.001.0001

Library of Congress Cataloging-in-Publication Data
Simmons, Dana (Dana Jean), author.
Vital minimum : need, science, and politics in modern
France / Dana Simmons.
pages cm
Includes bibliographical references and index.
ISBN 978-0-226-25156-1 (cloth : alkaline paper) — ISBN 0-226-25156-X
(cloth : alkaline paper) — ISBN 978-0-226-25173-8 (e-book) — ISBN 0-226-25173-X
(e-book) 1. Basic needs—France—History—19th century. 2. Basic needs—
France—History—20th century. 3. France—Social conditions—19th century.
4. France—Social conditions—20th century. I. Title.
HC275.S57 2015
306.0944—dc23
2014042626

♾ This paper meets the requirements of ANSI/NISO Z39.48-1992
(Permanence of Paper).

CONTENTS

CHAPTER 1

Introduction

"The fantastic part of a global recession," wrote one observer in early 2009, "is that it makes us realize what we need, what we want and what we have already. In reflection, we can begin to understand the way in which our 'needs' and 'wants' influence our quality of life and our impact on the environment."[1] In the wake of the Great Recession of 2008–9, many commentators wondered, "What do we really need?"[2] Pay cuts and layoffs forced Western consumers to spend less, and some saw an opportunity for us to reconsider what we buy and use. Proponents of "smaller living" stressed the pleasures of relationships and experiences over the fleeting satisfactions of material goods. "I think many of these changes are permanent changes," mused an economics expert. "I think people are realizing they don't need what they had."[3] Cured of consumerism, we might discover our true selves and our "real needs."

What is a need; what is a want, a desire, a luxury? To posit a need is always a political act. It means defining what is universal and law-like, and what is contingent; what is common to all people or all animals, and what is relative to different genders, classes, cultures, or races. The right to a minimal standard of living is a basic tenet of welfare and international human rights law.[4] But needs also serve to reinforce social difference. The so-called nature of sex or race, one group's inferior needs, has justified lower wages and nutritional standards for women and colonial populations.[5]

The politics of human need were central to the rise of the European welfare state. Europeans in the nineteenth and twentieth centuries experienced radical transformations in the state, economy, and family. Through measures of need, people from all walks of life disputed the conditions of wage labor and the modern social order. Scientific experts presented themselves as political authorities and claimed that the laws of society should strive to reflect those of nature. In the wake of the French Revolution, the logic of customary law and divine right no longer sufficed to explain social inequality. Social institutions were judged by the extent to which they fulfilled or violated essential human needs.

To imagine life in terms of needs is to imagine the parameters of human possibility. Needs express a set of ideas about life and its limits. Needs also refer to material things, food, clothing, housing, and other consumer goods. Needs mobilize technologies of measurement—scale balances, calorimeters, social surveys, statistics; and technologies of wages and welfare—minimum wages, family allowances, poor-aid and welfare programs. This book is about the technopolitics of human need.

*

On November 17, 1790, as the tricolor Revolutionary flag rose on public buildings across Paris, Antoine Lavoisier presented a "Memoir on Animal Respiration" to the French Academy of Science.[6] Lavoisier owed much to the Old Regime, notably the fortune he had accrued as a guarantor of the royal tax collection. He had much to lose when the Assemblée Constituante abolished the feudal order and the institution of tax farming, from which he drew his income. Yet Lavoisier was deeply imbued with an Enlightenment spirit of scientific rationality and public service. He joined those who sought a new logic of social organization, grounded in nature and reason rather than in privilege and power. His memoir on respiration was a seminal document in the history of modern chemistry; it was also a manifesto for a new society.

Animals, Lavoisier suggested to the Academy of Science, were "combustible bodies" like a lighted candle or an oil lamp. Just as a lamp slowly burns up its fuel, animals "burn and consume themselves" in the process of respiration.[7] Animals' fuel came from their own bodies in the form of digested food. Lavoisier himself had burned many candles in his experiments on the chemistry of heat. He understood combustion as an exchange of matter, which could be observed and measured. These experiments then became a model for his thinking about animals. Candle wax and oil provided "nourishment" for combustion in the same way that food gave substance to animal respiration. Both were subject to a delicate balance.[8] Animals must constantly replenish the matter that they burn or, like a flame, they expire.[9]

Lavoisier claimed that his research would uncover "almost every part of the animal economy."[10] Animals, he explained, were in continuous exchange with their environment. They absorbed matter through ingestion and inhalation, and they released heat and air in return. The flow of matter through the animal system, like a chemical reaction or an economic transaction, should result in a perfect equation. All that goes out must be replaced.[11] This was Lavoisier's principle of the balance, which regulated all his activities both worldly and scientific.[12] As a tax collector and a chemist he employed one of the most precise scale balances available at the time. He expressed his scientific ideas in the form of numerical equations. He deeply believed that all things, even moral and social things, were commensurable. Animals, people, and objects could be measured in terms of inputs and outputs. Essential life processes, nutrition and respiration, were the basis of Lavoisier's animal economy.[13]

Nourishment and economy were hardly academic questions in 1790. Food—bread—was the rallying point of the Revolutionary Days of October 1789, which forced Louis XVI to submit to the authority of the Assemblée Nationale. In these early days of the Revolution, legislators and protesters imagined that Old Regime charity would give way to a new regime of economic rights for poor citizens. A faction of the Assembly submitted that a universal "right to subsistence" be included in the Declaration of the Rights of Man.[14] Lavoisier confronted the political implications of his theory head-on. Just as he had inferred the workings of the animal economy from close observation of a candle, he applied his scientific method to society. Much of Lavoisier's speech to the Académie sounded less like a scientific report than like a political tract. He exposed a fundamental tension between human needs and social inequality.

> As long as we consider respiration only in terms of air consumption, rich and poor share the same fate; air belongs to everyone and costs nothing. . . . But now that experience has taught us that respiration is truly a combustion that consumes a portion of the individual's substance at every moment . . . [and] that this consumption increases proportionally when an individual leads a more active and laborious life, a whole set of moral considerations emerge from these physical results. . . . By what fate does the poor man, who lives by his hands' work and must use all of the force that nature has given him to earn his subsistence, consume more than the idle man, who needs less to repair himself? Why, in a shocking contrast, does the rich man enjoy an abundance that he does not physically need, when this abundance would seem intended for the working man?[15]

Lavoisier's respiratory experiments led him to identify a disjunction between nature's balance, wherein men consume what they need, and existing social inequality. He shared the Enlightenment belief that in the absence of social impediments, the forces of nature would find their optimal equilibrium. Yet his conclusions pointed to a schism between physiological needs and the social allocation of goods.

Lavoisier's memoir set out the problem of human need, at the intersection of science and politics. The Revolution gave this question urgency. The events of 1789 appeared to evacuate the authority of custom, monarchy, and divine right. Citizenship overtook subjecthood; individual rights overturned collective customs. The new society required a foundation in universal reason and human nature. For Lavoisier and others imbued in the liberal values of the Enlightenment, social institutions should be deduced from nature's first principles. In place of the traditional modes of structuring the self and society, they turned to science. As Lavoisier put it, "A whole set of moral considerations emerge from [science's] physical results."[16] In his memoir, scientific knowledge appeared as a standard by which to measure and guide the social order. Natural knowledge made social injustice apparent.

The imbalance that troubled Lavoisier would haunt the nineteenth and twentieth centuries. His speech to the Académie des Sciences offers a window into the central problematics that I will develop in the following chapters.

*

At the dawn of the Long Depression of the nineteenth century, a common figure emerged in France, Germany, Great Britain and the United States: *minimum vital*, *Existenzminimum*, *living wage*. Each of those terms signified a measure of life, below which the working classes could and should not fall. For American and European labor movements, this minimum represented a strategy to deal with the inexorable spread of wage labor. Instead of denouncing wages as "slavery," workers claimed a living wage sufficient to preserve their dignity and their rights.[17] "The first thing . . . trade societies should settle is a minimum," wrote British radical Hugh Lloyd Jones in 1874, "which they should regard as a point below which they should never go. . . . Not a miserable allowance to starve on, but living wages."[18] The notion of a "vital minimum," a living or "necessary" wage, drew from a scientific and political history going back to Lavoisier's age. It emerged from an intersection of socialism, political economy, physiology, hygiene, and population politics. In this book

I am interested in the meaning and function of "life" in living wage, the "vital" in vital minimum.

My argument is that a science of human needs undergirded the modern wage economy and the welfare state. A significant current of modern French human sciences—which engaged agronomists, chemists, doctors, anthropologists, economists, sociologists, amateur data gatherers, and trade unions—sought to measure human needs. These actors by and large did not trust in a providential economy to guarantee the reproduction of labor and the social order. Instead they believed that social organization, and particularly the circulation of goods, should be directed according to scientific principles. They came to this position from a variety of ideological standpoints, from reactionary conservative to revolutionary socialist. They had in common an antiliberal, bureaucratic orientation. Through measures of human need, scientists, workers, and others articulated a politics of life itself.[19]

Economies of need appeared to govern all biological and social processes. Anatomists believed that an animal's physical structure could be derived from its basic needs. Physiological features corresponded to needs; each organ allowed an animal to meet a specific need. Ethnographers measured a people's level of civilization along a scale of "primitive" and "advanced" human needs. Continental political economists "placed at the center of their analyses the question of human needs and their satisfaction."[20]

Above all, needs forced people to work. Economists and socialists alike argued that workers were compelled to labor in order to fulfill their needs. Pierre Proudhon put it succinctly: "The need for subsistence drives us to industry and work."[21] This formula, labor as a necessary means to subsistence, may be considered "the fundamental legitimating ideology of the capitalist social formation as a whole."[22] Wage work appeared natural, in accordance with the laws of physiology and human nature.

The scientific problem of subsistence, as Lavoisier and many others saw it, was how to convert workers' wages into life. The modern sciences of subsistence—including biochemistry, physiology, nutrition, and social surveys—were implicitly sciences of wage setting. These sciences employed a transactional model. They studied isolated individual subjects, not large groups or populations. They measured exchanges, the circulation of matter and goods, between individuals and their environments. They employed techniques of financial accounting, precisely recording flows of inputs and outputs and often converting their measurements into monetary terms. Life, in this view, was a transaction between an individual and his environment.

Scientists of subsistence studied the conversion of inputs into outputs, work into needs, needs into work. They sought to establish an equation of work and life. Most of these studies were directed toward governments and employers; despite this, many workers read them and responded. Workers' journals and unions complained that scientific studies of family budgets and nutrition masked social inequality. If we view subsistence as an individual transaction, they complained, then we cannot perceive the conditions that distribute wealth and well-being unjustly and unequally. Workers argued that only statistics and social surveys of family budgets could capture needs as they expanded over time. The notion of a vital minimum grew out of exchanges between scientists of subsistence and labor activists.

The "life" in living wage was a pure *bios,* what we might call an abstract life. The measure of needs, in a living wage, defined the cost of labor power. "Life" did not refer to any one existing person and was not an individual prescription in a medical sense. (In fact, French doctors had relatively little to do with measuring needs.) This was the same kind of life that occupied standardized, uniform rental housing units and performed rationalized tasks under the division of labor. Perhaps it may have been the same kind of life for which geographers and demographers later claimed "living space," *Lebensraum*, or *espace vital* in the colonies.

In another sense, however, the living wage was very concrete. It described the goods that workers made and purchased for themselves. As the list of commonly accepted basic needs grew over time, the market for new products grew alongside it. Needs standards tracked the supply and demand for consumer goods.[23] The living wage set a horizon for workers' desires, aspirations, and quality of life.

This form of life, the "vital" in vital minimum, was a very peculiar kind of life. The inverse of the life in living wage was not death. Eighteenth-century political economists Thomas Malthus and David Ricardo may have imagined such an opposition between wages and death. They posited that wage levels would bottom out at a socially necessary level when the working population began to starve and have fewer children.[24] In practice the living wage had little to do with that kind of mechanical and morbid view, which for Ricardo in any case was purely theoretical.[25] Minimum standards of need never measured the limit between life and death. This was not the "bare life," the "life exposed to death" that Giorgio Agamben describes as the "originary political element."[26]

The opposite of the "vital" in vital minimum was not death but unproductivity. To live, in this sense, was to be useful, to function, to produce and reproduce. A body's functions, the composition and work of its organs, responded to its needs. Needs defined what was necessary to *do*. A living wage produced use value. It supplied the means for labor to reproduce, that is, for work to start again the next day and the next year. For many, the vital minimum included two elements: one to repair a worker's exhausted body and another to allow families to raise children to take over their parents' work. Those who did not contribute toward this use value (by working for wages or having children) did not deserve the vital minimum. Those people were, in the words of Parisians under siege during the Prussian War, "useless mouths." The "life" in living wage was not an antonym of death. Life meant the ability to produce and reproduce.

In this book I follow this form of life as it traverses frontiers of discipline, class, gender, race, and power. I draw inspiration from historians of science who pursue keywords, foundational terms of modern existence, with a willful disregard for the fraught and useless opposition of "science" and "society."[27] François Vatin did so for "work," Anson Rabinbach for "fatigue," Margaret Schabas and Timothy Mitchell for "economy," John Tresch for "machine," Ludmilla Jordanova and Michelle Murphy for "reproduction."[28] Here I pursue the historical construction of the minimum wage, or, more abstractly, the equation of life and need.

"Vital minimum," "minimum existence," and the "living wage" expressed a social necessity—the need to reproduce labor—in the language of life. Living wages measured what Karl Marx called "prime necessities, naturally and historically developed."[29] The minimum wage was and is a site of constant struggle over life, production, and reproduction. The struggle was and is only ever resolved temporarily, by the assertion of particular forms of measurement and techniques of power. In post-Revolutionary France, social difference no longer appeared to derive from one's position in a customary hierarchy. A new logic for explaining class distinction had to be found.[30] Jean-Claude Perrot identifies this as the beginning of the "social question" which became central to nineteenth-century political thought. "The natural composition of private interests ceased to be a postulate and became a question. At this point the social survey [and the human sciences more generally] became irreplaceable."[31] Inequality had to be grounded in the nature of society, as revealed by scientific observation. The nature of need was not the same for all people; each age, sex, race, temperament, class, and profession had its own animal economy, which the scientist's task was to discern.

The dilemma that Lavoisier identified—the imbalance between workers' physical needs and the poverty of their means—was not resolved by the French Revolution. The spread of wage work across the nineteenth century made questions of subsistence ever more urgent. Wage workers, from the 1830s to the early twentieth century, claimed a basic "right to live."[32] Following the 1848 revolution, the Second Republic echoed these words when it "committed to providing for the worker's existence through work."[33] Wage debates in the nineteenth century, like twentieth-century politics of welfare, hinged on survival, existence, security, and well-being.

Economic life underwent major transformation over the course of the nineteenth and twentieth centuries, with the spread of the wage form. Guild and family economies gave way to wage transactions. Half of the active French population in 1850 sold its labor for wages, and almost all wage earners were manual laborers.[34] The nineteenth century was "the century of putting to work."[35] A second major wave occurred in the generation after the Second World War; by 1975 over four-fifths of French citizens worked for wages.[36] Wage contracts engaged individuals as autonomous workers, not as members of any social unit.

The price of a laborer's wages, in the classical economic view, was set by the free market. Yet that price contained no guarantee that workers would be able to maintain themselves. For an industrial economy to thrive, workers must not only replenish their own bodily substance; they also must raise children to replace themselves in the future. Labor must be constantly renewed, that is reproduced. How could one allow the market to set the price of wages and yet ensure the reproduction of labor? There seemed to be no way to reconcile the conditions of industrial work with the needs of a worker and his family. This dilemma, and the search for a scientific solution, defined the social question central to the modern human sciences.[37] The equation of needs and wages was—and remains—a critical problem for social thinkers in the modern era. In it lay the crux of the social question and, to some, the potential to resolve class conflict.

The question of needs was deeply gendered. Historians of gender and the welfare state, after Susan Pedersen, distinguish modern France for its paternalistic emphasis on the family as a site of social intervention.[38] This book examines the social-scientific roots of the French emphasis on families. Social surveyors in the mid-nineteenth century employed a normative model of a unitary nuclear family. This became their fundamental unit of analysis, to which all other social differences were subordinated. Differences of power, resources, or needs within families—such as the unequal distribution of wages or food among family members—did not factor in their consideration. In nineteenth-century social science, women and children appeared almost exclusively as indivisible parts of a family unit. Women's own needs entered the picture only in the form of a threat to family integrity and the natural social order. Twentieth-century "paternal" welfare institutions drew upon this social-scientific legacy by targeting the family as the sole legitimate recipient of social aid.

Men, by contrast, did appear autonomous from the family in their public role as "the worker" (regardless of the relatively high rate of women's work in this period). The worker's body appeared both masculine and self-contained. Female bodies were strikingly absent from physiological studies of labor and nutrition until well into the twentieth century. Nineteenth-century agronomists went so far as to posit a model of sexless reproduction of labor. In an analogous occlusion, nineteenth-century liberal economists considered men the only legitimate and natural recipients of wages. Only men could assure subsistence and reproduction. Women's work, by contrast, was considered supplemental, superfluous, excessive.[39]

Nature itself allocated unequal needs to different individuals. If women, the poor, and colonial subjects were judged by science to have lesser needs, their inferiority appeared necessary and natural. "Within [a] revolutionary liberal framework, an appeal to natural rights could be countered only by proof of natural inequalities."[40] Social difference—of race, class, or gender—was made necessary in this process. Difference was concretized in the inequality of natural needs.

Needs, hunger, and poverty were "core dilemmas" of the modern welfare state, dilemmas "that helped determine where the boundaries would be drawn between the market and the state, the subject and the citizen, the individual and the collective, the nation and the empire."[41] Historians have shown that political economy and "market culture" suffused modern social surveys and public health measures.[42] I argue that the modern sciences of subsistence, nutrition and consumption were implicitly about wages and the distribution of wealth.

*

The opening chapters of this volume examine two models of the nature of need: the agronomic and the anthropological. A group of chemists and agronomists under the July Monarchy took up Lavoisier's search for scientific solutions to social inequality. In chapter 2 of this volume I describe the lasting influence of chemistry on French science and politics. These scientists were deeply engaged in politics and shared widespread concerns about the potential for popular unrest. They applied their expertise to the most pressing issue of the day: the condition of the working classes, particularly poverty and poor nutrition. One weighed and analyzed animal excrement in order to set a minimum "maintenance ration." Another calculated daily wages for his farmhands based on the amount of wheat required to fuel different kinds of work. As chapter 3 shows, these agronomic measures were applied extensively by state officials under the July Monarchy and the Second Empire. Scientific commissions and state administrators weighed and regulated the diets of prisoners, schoolchildren, hospital patients, and charity cases. All citizens, following Lavoisier's principle, were subject to the scale balance.

Social surveyors of the mid-nineteenth century, like agronomists, sought to establish natural measures of need. In chapter 4 I trace the origins of the nineteenth-century social survey, in a collision of biology and politics. Social surveys drew ideas and methods from anatomy, anthropological medicine, economic theory, scientific exploration, and worker activism. Common to all of those fields was an obsession with measuring and categorizing human needs. Anatomists Georges Cuvier and Jean Baptiste Lamarck defined modern zoology in a debate over the role of needs in the animal economy. The zoological conflict then profoundly influenced theories of social welfare. Zoologists and social reformers alike asked whether needs were fixed or mutable. Conservative social reformers Joseph de Gérando and Frédéric Le Play divided the French population into fixed sociomedical, then racial, types. Socialists and workers' journals argued for a statistical index that could track ever-expanding needs. Need appeared as a foundational category for class politics and for the natural and human sciences.

The second half of this book describes the history of the "vital minimum" from the nineteenth-century Long Depression through the Fourth Republic. Despite three epochal ruptures between the 1870s and the 1950s—the rise and fall of the Third Republic and the caesura of 1940–44—the vital minimum traversed a century of French history. The vital minimum was the cornerstone of wage and welfare policy across the Third Republic, Vichy, and the Fourth Republic. It became so engrained in French social imaginary that even after the government officially disavowed the vital minimum in the 1950s, citizens continued to ask government officials about the "vital minimum wage."

Chapter 5 suggests that food rationing was an important precursor to the national minimum wage. During the Siege of Paris (1870–71), the Parisian provisional government imposed rationing and proclaimed itself as the guarantor of its citizens' well-being. The siege dramatically displayed the disruptive potential of needs under conditions of scarcity. Only some people—the active male citizens on whose existence the country's survival depended—could reclaim satisfaction of their needs. Parisians distinguished between citizen-soldiers, who deserved a minimum wage for their service, and those they named "useless mouths." This distinction became a matter of survival as the state took on the responsibility for distributing dwindling resources. Usefulness to the state became the standard for needs and rights. This paved the way for the development of the welfare state in the following decades.

Wage debates in twentieth-century Europe were dominated by a new regulating principle: the *minimum vital*, "living wage," or *Existenzminimum*. In chapter 6 I analyze the political-scientific content of the "vital" or "living" component of the living wage. The figure of the vital minimum traveled from classical political economy to physical energetics, Marxism, and social Catholicism, to social statistics, to Fordist wage regulation and welfare. Welfare politics invoked both physiology and sociology via the vital minimum. The French welfare state first mobilized the vital minimum as an individual minimum wage. During the First World War the vital minimum developed into a collective family allowance. Chapter 7 describes the central role of needs measures in occupied France from 1940 to 1944. A group of antiliberal ideologues, engineers, and economists shaped the Vichy regime's social policy. These men propounded a "science of man," designed to overcome class conflict. In this chapter I develop two case studies, food rationing and the "vital" minimum wage. I outline the scientific vision behind the authoritarian welfare state of Vichy France: a hybrid of neoclassical economic theory and biosociology. The twentieth-century European welfare state was as much a racial-eugenic regime as a social-democratic one. In chapter 8 I follow the fraught and occasionally comic deliberations of a committee of unions, employers, and experts appointed in 1951 to fix a "model" minimum wage (known as the SMIG). The SMIG commission enacted a political contest between workers, employers, and the state. As it called upon expert testimony, the commission participated in the emergence of an empirical, policy-oriented postwar social science. More broadly, the commission's work reflected French citizens' everyday struggle to reconcile scarce resources and expanding consumer desires. Through debates on meat, coffee, and tap water, commissioners grappled with the constraints and desires of common citizens. The SMIG shows us how entangled were class politics, science, and consumption in the postwar French welfare state. Ultimately, the SMIG project proved a failure. Employers, unions, and finally the Fourth Republic government repudiated the very principle of the vital minimum.

The French welfare system distinguished family needs from workers' wages, charity payments from salaries. Many believed that this regime would shelter citizens from capitalist excesses, while leaving wages to float on the free market.

The system depended on a clear demarcation between needs and desires, welfare and the market. Yet no one could agree on a list of goods and services that defined basic needs. Should one count coffee or a vacation at the beach? What about union membership or toilets? The epistemological crisis spilled into the political crisis. In the concluding chapter I consider the broader meaning of this spillage from natural to social, epistemological to political.

In this book I elaborate encounters between multiple natural-scientific and social-scientific fields and welfare institutions under the rubric of minimum needs. Such a project cannot offer a continuous or encyclopedic account of individual disciplines or institutions. To do so would result in a ponderous and unmanageable compendium. Instead I opt for detailed studies of key points of intersection, in which a diverse range of actors work through the science and politics of need. I trace passages between fields of knowledge and political activity, transactions that have not been systematically studied.

I do not presume from the outset to write a history of economics, sociology, physiology, gender, labor, or the welfare state. Connections between these fields, and outside them, create a far more complex and interesting picture than one bounded by contemporary disciplinary strictures. Had I stuck too closely to the latter I would have missed the central role of agronomists, who in the 1840s articulated a science and politics of need. The arcane deliberations of an obscure 1950s subcommittee on minimum wages might not have appeared so rich in contradiction and significance. This is a history "intertwining . . . the cognitive, the material, the social and the normative."[43] I have tried, to the extent of my power, to retain a touch of the concrete and quotidian stakes involved. At issue were daily budgets and expenditures, whether families ate meat or bread, whether homes had access to heat or running water. There were real consequences for poor and working-class men, women, and children. Their voices filter—more rarely, explode—through the medium of pamphlets, official reports, and social surveys. I have sought to keep this work open to multiple registers of language, measurement, and meaning.

CHAPTER 2

Subsistence

We begin our story with two agronomists and an insurrection. Adrien de Gasparin, a "vigorous man, robust and very ugly . . . with a satyr's physiognomy, intelligent and gay," arrived in Lyons in November 1833.[1] A former military officer, expert on veterinary medicine and author of a treatise on "agriculture and government," he had just been appointed prefect of the most volatile region in France. Lyons was a political powder keg. Silk weavers fought against wholesale merchants over the price of their piecework; at stake were thousands of livelihoods and a backbone of the city's economy. In 1831 silk weavers demanded a minimum pay scale. When merchants refused to sign on, workers called a strike and set up barricades, brandishing the motto, "To live by working or die fighting."[2] The July Monarchy crushed the 1831 uprising with full force. Gasparin, then prefect of the neighboring Isère region, mobilized three thousand men to march into Lyons and help military troops to restore order.[3] Thanks to this display of authoritarian zeal, he was transferred there two years later.[4]

As prefect of Lyons Gasparin dealt harshly with weavers' renewed requests for wage controls. He claimed (it is said) that government should not interfere in the "free" transactions between capitalists and workers.[5] When workers went on strike in 1834 over a pay cut, Gasparin had the strike's leaders arrested and tried in court. Again, barricades arose in the city's poorer neighborhoods; this time, they were knocked down by cannon fire and ten thousand soldiers' bayonets.[6] The insurrections of 1831–34 deeply scarred the city and the country. Historians trace the origins of French working-class political identity to that moment.[7] Gasparin himself received the highest honors that a grateful king, Louis Philippe, could bestow, including a title of nobility.[8]

Within months, traces of the 1834 insurrection were effaced. The French state inaugurated a new university on the same square (the Place des Terreaux) where striking workers had assembled in protest. A very different kind of crowd assembled in January 1835 to hear an inaugural address by the dean of the new Faculty of the Sciences, Jean Baptiste Boussingault. Boussingault was fresh off the boat from South America, where he had served for ten years as scientific attaché to Simón Bolívar. He was less than thrilled to find himself in a provincial industrial town with no academic tradition to speak of. On the university's opening day Boussingault made a rousing case for an applied, pragmatic, and politically infused science, a science in the national interest. He presented prospective students with a vision of sugar beets, spinning jennies, and chemical applications that could improve the lives of the common people.

In sending Boussingault to Lyons, authorities must have hoped that bourgeois science would redeem social crisis. But Lyons's bourgeois citizens did not receive his speech well; they must have been in no mood for unity and conciliation.[9] Boussingault lasted only a few months in Lyons. He moved away at the first opportunity and set up the chemical laboratory in which he would perform his most important agronomic experiments.

Two agronomists intervened at the margins of the first modern minimum wage strike in France. This conjunction was not as incongruous as it may seem. At issue was the problem of subsistence. What, and how much, did a worker need to live? Who would measure, who would guarantee, each citizen's daily ration? In a world where people worked for wages and bought their food in the market, those who needed the most food often received the least. This was Lavoisier's dilemma, given urgency by an expanding commercial economy.

I argue below that agronomists were perhaps the most important political economists at work in the mid-nineteenth century.[10] Agronomists pioneered a theory and practice of wage management. Years ago, François Dagognet forcefully argued that nineteenth-century agronomy marked the debut of a modern rationalized, industrial bioscience.[11] The implications of his argument have not been fully explored, particularly by those interested in social theory. As economic historian Mary Morgan remarks, agriculture "lay at the heart of the two main problems of political economy" in the modern period: population and consumption.[12] In the following pages I describe key technologies and social formations, which placed agronomists at the center of the modern wage question.

The mid-nineteenth century was an era of "agromania." Scientific agriculture spread rapidly in this period, and dozens of agricultural manuals rolled off the presses.[13] Agricultural productivity bore directly on a nation's prosperity and growth. Everywhere in France, as the surface area devoted to agriculture expanded, grain grown for the market edged out other crops.[14] The majority of agricultural produce in this period shifted from home consumption to sale, from subsistence to commerce.[15] Social reformers such as Louis René Villermé idealized rural life and the agricultural economy.[16] Agronomists grappled with questions of work, subsistence, class, and the distribution of wealth.

Agronomy bridged the life sciences, natural sciences, and social policy. Like nineteenth-century physics, concerned with measuring work and waste, agronomy was an instrumental science. Plants and animals appeared as machines, whose inputs and outputs could be measured and adjusted. Living machines operated within a global flow of matter. "Chickens, sows and cows little by little became micro-factories, tied to other, larger, factories . . . a new space emerged, of agro-industry."[17] The scientists in this chapter employed chemical methods to answer physiological questions. They were perhaps, as Dagognet claims, the first biochemists.[18] Agronomists in Lavoisier's day studied the productivity of entire fields or regions. These scientists, by contrast, oriented their research toward the individual organism and its interactions with its environment. Nineteenth-century agronomists viewed plants and animals as objects of economic management, whose needs could be measured in units of carbon and nitrogen.

The agronomists under study in this chapter—Jean Baptiste Boussingault and his mentor, chemist Jean Baptiste Dumas—gained scientific and political prominence under the July Monarchy (1830–48) and the Second Republic (1848–51). Both held more than one government post, both served on numerous state-appointed scientific commissions, and both were elected to national academies of science. Though their political party affiliations diverged, they shared a common faith in technical solutions to the economic and social problems of their day.

If one were to insert these agronomists into John Pickstone's grid of nineteenth-century science and ideology, one might call their scientific and social vision "romantic-bureaucratic."[19] They conceived of the universe as a great circulating flow of matter. The whole cycle was far more important than individual bodies, which appeared simply as stopping points for the flow of chemical elements. In this model individual animal bodies did not alter the matter passing through their systems, but simply used and ejected what they took in. Dumas and Boussingault believed that animal bodies were unable to transform matter (such as sugar to fat), which led to a fractious debate with their contemporary Justus von Liebig. This subordination of the parts to the whole made their agronomic model attractive to romantic anticapitalists.[20] Dumas and Boussingault themselves were deeply implicated in the market economy. They sought to use their chemical model as a stabilizing force to direct the flows of matter, goods, and capital.

Scientists of agriculture sought out efficiencies, quantified and optimized resources. As Jean-Antoine Chaptal wrote in his *Chemistry Applied to Agriculture* of 1823, "In agriculture, as in all well conducted business, everything is calculation; and the operations of agriculture should be ruled by expenses and receipts." Calculation, in this context, was financial, and it was aimed to reduce expenses to a minimum. "Agriculture requires only what is necessary, and it reproves of excesses as a kind of luxury."[21] Even as agronomists adopted scientific methods drawn from chemistry and medicine, their objectives remained pragmatic. The success of experimental results relied on accurate measurement and financial accounting; did a new source of manure or feedstock maintain or increase the output of plants and animals?

Their measures of input and output, carbon and nitrogen, became standards for measures of human needs. When the World Health Organization releases guidelines for minimum protein and carbohydrate intake, when food labels list recommended daily allowances, when school canteens weigh their meal portions, these operations are inherited from nineteenth-century agronomic science.

PIGS ON A BALANCE

For 205 days in 1844, Boussingault imprisoned his pigs. He locked them in individual cells, deprived them of their habitual straw bed, and obliged them to take their rest on hard wooden boards. "The two sequestered pigs," he reports, "rarely exited [their cells] and then only for bathing." The cells contained small peepholes "to obviate the serious disadvantages that a rigorous regime of solitary confinement would certainly present." For the duration of their confinement, he carefully gathered, weighed, and analyzed the chemical components of the pigs' feed and excrement. He compared the nitrogen and carbon ingested with that lost in combustion or rendered by bodily secretions. His animal prisoners served to establish what he called "the nutritive value of a daily ration."[22] Boussingault's work on animal feed focused on the balance between inputs and outputs, ingestion and excretion. Drawing upon Lavoisier's legacy, Boussingault viewed nutrition as a process of expenditure and replacement. "An animal subjected to the maintenance ration," he wrote, "renders, in the various products resulting from its vital action, a quantity of organic material precisely equal and similar to the amount that it collects in its food."[23] He invoked a chemical balance to equate consumption and expenditures. "It is fairly certain that . . . one should find in [an animal's] excretions, secretions and the products of its respiratory organs, the totality of all elementary matter contained in the foods [that the animal] consumed."[24] In each of his feeding experiments he produced tables recording the weights of chemical elements that flowed into and out of his animals. Boussingault and his mentor Dumas elaborated on Antoine Lavoisier's science of consumption.

Lavoisier formulated his famous dictum that "nothing is created; nothing is lost" with reference to organic reactions.[25] He described bodily processes as a continuous absorption, combustion, and ejection of matter.[26] Life could be accounted for with a chemical and economic balance sheet.[27] Each day, Lavoisier wrote in 1792, "a sum of matter equal to what is received in the intestinal canal is consumed and expended, whether by transpiration, by respiration or finally, by the different excretions."[28] Instrumental measurements, for Lavoisier, were the only means to access natural phenomena. Measures made natural causes knowable.[29]

In 1793 Lavoisier published a revised version of the *Memoir on Respiration*, which he had read to the French Academy of Science at the start of the Revolution. It would be one of his last major works. Lavoisier believed that human needs, as revealed by science, would provide a measure and a guide for revolutionary society. Nature, rather than divine law, tradition, or popular demands, would provide a steady principle of governance. With this *Memoir* Lavoisier posited human need as a scientific problem and as the key to a scientific social order. Lavoisier's 1793 memoir on respiration is recognized by historians of science as a seminal document in the modern chemical revolution. It was that and more: a political tract for a new society.

In nature, Lavoisier submitted, human bodies display an astonishing ability to vary their consumption to match vastly different conditions of exertion, temperature, and climate. "At each moment in the animal economy one can observe a truly amazing result: continuously variable forces, continuously in equilibrium, which allow the individual to adjust to any circumstance that he may encounter."[30] Nature appeared prior to social institutions and superior to them.

> The physical order, subject [*assujeti*] to immutable laws, has achieved a state of equilibrium that nothing can disturb; the physical order does not undergo the tumultuous movements that one sometimes sees in the moral order.[31]

"Let us not vilify Nature," Lavoisier tempered, for the outrageous fact that poor laborers work far harder yet eat far less than their rich and idle neighbors. Instead "let us blame the defects inherent in our social institutions."[32] Inequality was not the work of nature but of man.

Lavoisier came to understand the exchange of airs and nutrients in animal bodies as a form of consumption. His memoir equated respiration and combustion: "From this point of view animals that respire are veritable lamps that burn and consume themselves."[33] The analogy between the body and an oil lamp was dual: air provided oxygen and heat for the fire, and nutrients in the blood, the "very substance of the body," formed the fuel to be burned up. This model, I would argue, established a foundation for a modern science of consumption.

Lavoisier's early work on respiration employed a terminology of "alteration," drawing an analogy between animal and physical systems. He treated common air as a compound substance, containing measurable quantities of vital gases. Once he had determined the "just proportion of vital air and *azote* gas" that permits animals to breathe, he set about measuring deviations from that proportion.[34] In particular he warned against the "mephitic air" that tended to form in crowded public spaces. By 1790 his language shifted: "alteration" and "just proportion" gave way to observations on the "consumption of vital air."[35]

The language of consumption allowed Lavoisier to apply the combustion model to human labor. In his memoir Lavoisier suggested that one could classify various types of work according to a common physical measure: the byproducts of bodily combustion. He installed his assistant in a breathing apparatus that allowed the young man to vary his rate of movement and exercise. The instrument permitted Lavoisier to store up and weigh the matter exhaled under different conditions of human work and rest. Through this method, he claimed, he could compare activities that otherwise appear to have nothing in common. The efforts expended by a musician playing an instrument, a philosopher deep in reflection, a politician giving a speech, and a brick hauler could all be reduced to their "mechanical" element.[36] Lavoisier measured the "mechanical" work of his subject in terms of a weight raised to a certain height. He converted all forms of exercise to that format and calculated the rate of respiration necessary to perform them. In this he referred to the physics of force and motion, which later became the science of work.[37] Lavoisier expressed work and need as two sides of a chemical-mechanical equation.

Needs, for Lavoisier, were law-like but not universal. "For each individual," he asserted, "there exists an undeniable law" for the scientist to uncover. Each

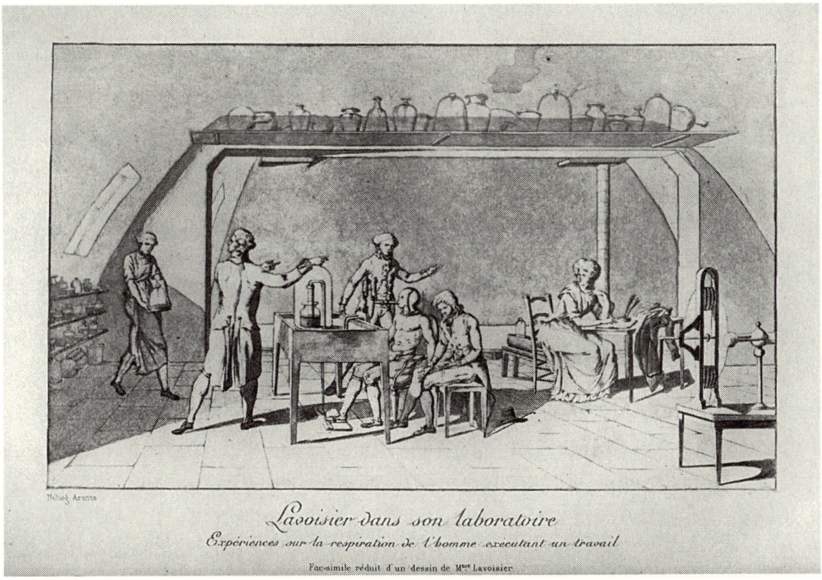

FIGURE 1. Lavoisier dans son laboratoire: Expériences sur la respiration de l'homme exécutant un travail. Fac-simile réduit d'un dessin de Mme. Lavoisier. Edgar Fahs Smith Collection, University of Pennsylvania Libraries.

person was subject to his or her own law, which was "quite constant" and could be measured.[38] "Consumption varies considerably in different individuals, according to their age, their state of vigor and their health, and whether they . . . are habituated to heavy work."[39] Lavoisier did not propose to set a single universal standard for nutritional requirements. Rather he suggested that the scientist must establish distinct measures for different people. This was the task of the patriotic physician whom he called to duty: "The physician, in the silence of his laboratory and his office, can also exercise a patriotic function; he can hope, by his work, to diminish the mass of ills that afflict the human species, to increase its happiness and its well-being. . . . The physician also can aspire to the title of glorious benefactor to humanity."[40]

Lavoisier's differentiated view of society led him to investigate individual consumers, classified according to their social status, at very close range. Like the physician's study of his patients' physical needs, Lavoisier the economist examined the innards of individual households. In 1791, at the urging of the Constituant Assembly tax commission, he assembled fifteen years' worth of research on the French economy into a summary volume entitled *Territorial Wealth of the Kingdom of France*.[41] Here he applied his principle of the balance to the French economy. "All that is consumed every year" on average, he wrote, "is reproduced every year."[42] To know how much France produces, one must measure what its population consumes. In devising such a measure Lavoisier was led to contemplate the import of social difference.

Lavoisier worked in the world of the French Enlightenment, a world where variations, accidents, and fluctuations were contained within stable boundaries. All natural systems, left on their own, would tend toward equilibrium.[43] Physicist Pierre Laplace constructed a model of the universe as a system fluctuating between stable boundaries. The seas and all the celestial spheres circulated from nadir to zenith according to a regular pattern. "As with all magnitudes subject to a maximum or a minimum," Laplace wrote in his *System of the World*, "the tide rises and falls towards its limits."[44] So did the sun, the moons of Jupiter, a ship's prow, the planets Mercury and Venus, and any system in movement. Old Regime reformer Anne-Robert Turgot observed "an equilibrium" in the French economy, "between the value of [agricultural] production, the consumption of different kinds of goods, different types of work, the number of workers and the price of their wages."[45] Mathematician Jean-Louis Lagrange designed a "calculus of nourishment" which tracked all foodstuffs produced and consumed in France using units of bread as an index value. He concluded that food prices fluctuate according to a good's nutritive value.[46] As the celestial bodies circulated in harmony between universal limit points, so did food, money, and people. Lavoisier inscribed his science of human needs within that tradition; the Revolution moved him to emphasize the chasm between a state of natural equilibrium and social inequity.

When societies ignored the natural laws of need, the result was poverty and death. The inadequacy of workers' wages threatened their physical health. Society accorded laborers, who used themselves up in work, fewer means to subsist than the idle rich. If a laborer did not receive enough nourishment to replace what he expended, Lavoisier warned, he eventually would waste away. At a certain point the body can no longer compensate for its lost matter and "this is when a diseased state appears."[47] Having described this physiological imbalance, Lavoisier set out to rectify it. He boldly proposed to transform the wage relation.

Wages, in his vision of the Revolutionary social order, would no longer be set by the market alone. Instead they would be fixed by a scientific scale of bodily expenditure. Science, he asserted, would lead the new society to "raise the price of work and ensure that it received its fair compensation."[48] He echoed popular demands for a "fair price" in an entirely modern scientific framework.[49] He all but anticipated Marx's dictum "to each according to his needs."[50] Instead of a language of rights, however, Lavoisier employed the terms of science. Physician-scientists, not politicians, were alone qualified to determine the level of need that wages should supply.

Lavoisier closed his 1793 *Memoir* on a note of cautious optimism. "Let us bless Philosophy and Humanity, which together promise to bring about wise institutions that will bring fortunes closer to equality, raise the price of work . . . and present to all classes of society—especially the indigent—greater pleasures and happiness." The union of philosophy and humanity, science and politics, appeared to offer an end to social strife. If, he tempered, the "enthusiasm and exaggeration" of the crowd did not "destroy the hopes of this country."[51] In 1794 Robespierre and the National Convention put Lavoisier, alongside all former tax farmers, to the guillotine.

Lavoisier's chemical calculus of need was roundly rejected by early nineteenth-century medical authorities. Experimental physiologists such as François Magendie turned to close observation of individual organs and rejected general formulas like the equation of inputs and outputs. The upheavals of the Terror had cast doubt on Lavoisier's confidence in science and social transformation. In the 1840s, however, an influential group of chemist-agronomists revived and refined Lavoisier's methods.

Boussingault and Dumas applied Lavoisian methods to the nineteenth-century life sciences. Dumas edited a collection of Lavoisier's collected writings and campaigned to revive interest in the Revolutionary chemist's work. Their approach followed Lavoisier's basic principle: "When someone supposed a transformation of substances to occur in the animal he should be able to explain it on the basis of the chemical compositions of the substances involved, reduced if possible to an equation."[52] In the 1830s and 1840s, chemists refined instruments that measured the relative quantities of basic elements in any volatilized substance. The theoretical framework of organic chemistry expanded alongside precision of measurement. French scientists adapted and improved Liebig's innovative *Kaliapparat*, which allowed chemists to produce simple and precise measurements of organic substances by combustion.[53] Quantitative organic analysis spread alongside Lavoisian chemical accounting.

SCARCITY

Lavoisier's dilemma—the imbalance of poor workers' needs and means—haunted the nineteenth century. The problem of subsistence posed a seemingly intractable challenge to the modern social order. When food supplies grew short in the nineteenth century, protesters in cities and villages invoked the

French Revolution. In rural areas, as in urban industrial zones, Robespierre's "right to exist" appeared everywhere as a rallying cry. Popular placards and brochures reclaimed revolutionary price controls on consumer goods. Some demanded state-enforced grain seizures and punishment of hoarders. Protesters laid claim to a Revolutionary right to subsistence.[54] In this context Boussingault and Dumas developed a scientific measure for minimum food rations.

Food focused and channeled broader political resentments and ambitions. Rumors of "famine plots," schemes by politicians and merchants to starve the poor, focused great energy still in industrializing nineteenth-century France. Hunger, warned Emile Zola and other observers, would lead to revolution and violence.[55] The cry "we are hungry" resounded at bread riots, strikes, and revolutionary actions. It united urban artisans and rural peasants, wage earners and farmers. Bands of women and men intercepted grain transports, disrupted marketplaces, and publicly remonstrated against public officials in the name of a right to bread.[56]

Severe economic crises shook France in 1828–32 and again in 1846–47, bringing hunger, scarcity, and social upheaval. In both cases, harvest failures—including the great potato blight of 1847—coincided with commercial and banking collapse. Industrial shocks struck both big manufacturers and artisans. Massive investment in railroad construction led to an overextension of monetary reserves.[57] Protectionist policies under the Restoration and the July Monarchy cushioned the impact of economic crisis for industrialists but made life even more difficult for small consumers. Prices for everyday goods soared.

In 1830 and again in 1848, popular disturbances over food prices coexisted with civil and political movements seeking higher wages, electoral reform, lower taxation, and revolution.[58] Both crises culminated in regime change, marking the start and end of the July Monarchy. The Glorious Days of 1830, which installed Louis Philippe as constitutional monarch, demonstrated the power of popular movements and recalled the Revolution of 1789. Perhaps even more disturbing to the ruling elite, because they could not be contained within the bounds of political regime change, were the Lyons uprisings of 1831 and 1834. State authorities were deeply perturbed by these events and the potential for class revolt that they portended.

Boussingault and Dumas had a strong sense of agronomy's social importance and were deeply involved in politics. Dumas spent years arguing for agricultural reform and public hygiene as Minister of Industry, Commerce, and Agriculture (1849), in the national Chamber of Deputies and Senate, and on the Parisian Municipal Council. As minister of the interior, Dumas established the first National Agronomic Institute at Versailles, a short-lived predecessor to the current institute in Paris; the institute's director, from its inception in 1850 until its closure in 1853, was Adrien de Gasparin.[59] In the final years of his career Dumas served as president of a successor Institute of Agriculture; Boussingault was its director of research.[60]

Dumas is best known for his work on the classification of organic molecules, and he held the world's first chair of organic chemistry at the Faculty of Medicine.[61] He taught for a time at the Ecole Polytechnique, the training ground for the country's technocratic elite, and was one of three founders of the Central School for Arts and Manufactures, dedicated to preparing civil engineers for work in industry.[62] By the 1840s Dumas was known as "the most powerful academic scientist in France."[63]

Both Dumas and Boussingault had middling backgrounds. Dumas apprenticed to a provincial pharmacist and a scientifically minded physician; Boussingault, son of a shopkeeper, became a mining engineer. Dumas married into a porcelain manufacturing family and leveraged his in-laws' wealth to establish a private laboratory and a network of personal scientific patronage.[64] Boussingault married the daughter of a wealthy Alsatian landowner. Both men rose to wealth and prominence through technical education, political connections, and lucky marriages. In large part, their scholarly and popular writings made them known, but their influence and prestige came from the state.

Boussingault entered politics on the wave of the 1848 revolution. He campaigned for the Chamber of Deputies with Louis Blanc's socialist party, though once seated he tended to side with the moderate republicans. He confirmed this moderation when appointed to the State Council in 1849. When Napoleon III dissolved the council in 1851 Boussingault returned permanently to the scientific realm, still under the wing of his powerful patron Dumas.

Boussingault had first encountered the problem of subsistence during a decade of scientific exploration in South America under the tutelage of Alexander von Humboldt. In the 1820s he managed a gold mine in the Colombian region of Antioquia, where he became interested in agronomy. The mine had no access to food producers; laborers had to return home every two weeks to supply themselves. "To stabilize the labor force their food supply had to be assured. It was then that we set up a large banana plantation [and] . . . encouraged planting of corn, yucca and root vegetables. . . . It was while organizing this tropical agriculture that I understood that we should demand from the soil the food necessary for the population; in a word, one must cultivate to live. My study of agriculture dates from that time."[65]

Following this question, Boussingault began to study the fertilizing effects of guano deposits on Colombian fields. His work with guano formed the basis for years of study on nitrogen fixation in plants and justified his acceptance into the Academy of Sciences. Through Dumas's patronage Boussingault obtained research and teaching positions at the Academy of Sciences, the National Conservatory of Arts and Trades (CNAM), and elsewhere.

Boussingault's family farm provided him with material for a life of research and, judging from his portly profile, plenty of subsistence. He applied his agronomic and engineering knowledge to manage the farm, which included crops, livestock, and a small petroleum mine. In a delicious conjuncture, Boussingault's innovations in agronomic chemistry took place next to the site of what would become Europe's first industrial petroleum extraction. (In his time, the petroleum was sold as lamp oil.) Laborers worked both the farm and the mine, which formed a single integrated agro-industrial site.[66] Boussingault's farm at Bechelbronn was identified by his contemporaries as one of the first agricultural research stations, preceding Rothamstead in England (1843) and Moekern in Germany (1852).[67]

Boussingault recounts that in 1840 drought hit Alsace, forcing farmers to seek alternative sources of feed for their animals. "At this time," he recalls, "we were obliged to replace a large portion of the hay typically consumed in our stables with potatoes; and, using [elementary chemical] analysis as the basis for this substitution, we obtained quite advantageous pecuniary results, while conserving the health and vigor of our teams."[68] This experience confirmed his passion for chemical analysis as a means to maximize crop yield and to minimize the quantity and cost of animal feed. He was well aware that the same techniques could potentially apply to people suffering from the effects of the drought.

Harvest failures in the 1840s brought social catastrophe. In January 1847, a crowd in central France joined in what one historian has labeled "the last great wave of food riots."[69] In the background was a severe subsistence crisis, during which food prices skyrocketed and wages and employment grew tighter. Entreaties for official price controls went for naught, and some residents literally took the sale of grain into their own hands. A passing convoy was seized, notables' houses looted, and one particularly distasteful lender murdered. Many French urbanites were shocked by news of the return of "jacqueries" in the mid-nineteenth century.[70] The workers' newspaper *L'Atelier*, on the other hand, had seen this coming. "We will tell the government," wrote the paper when the grain harvest first failed, "that a natural consequence of the rise of food prices [*subsistences*] must be a rise in wages. When [government] agents pitilessly beat up the workers who gather in order to rid themselves of the hunger that threatens them, we will give these acts the name that they deserve."[71]

FIGURE 2. Jean-Baptiste Boussingault (1801–1887), chemist and French agronomist. BOY-29160 1801. © Jacques Boyer / Roger-Viollet / The Image Works.

During these nineteenth-century crises, "traditional" grain protests merged with workers' demonstrations. Calls for fair food prices and equal access to the market joined with demands for higher pay. Price controls on essential goods were now rhetorically linked to similar controls over wage setting. The mayor of a village on the industrial periphery of the Choletais complained during the

TABLE 1 Jean Baptiste Boussingault, *Economie rurale considérée dans ses rapports avec la chimie, la physique, et la météorologie*, 3551

FODDER CONSUMED BY THE HORSE IN 24 HOURS

FODDER	WEIGHT wet	WEIGHT dry	ELEMENTARY MATTER in food: Carbon	Hydrogen	Oxygen	Azote	Salts & earths
	gr.	gr.	gr.	gr.	gr.	gr.	gr.
Hay	7500	6465	2961	323	2502	97	582
Oats	2270	1927	977	123	707	42	77
Water	16000						
Total	25770	8392	3938	446	3209	139	672

PRODUCTS VOIDED BY THE HORSE IN 24 HOURS

PRODUCTS	WEIGHT wet	WEIGHT dry	ELEMENTARY MATTER in products: Carbon	Hydrogen	Oxygen	Azote	Salts & earths
	gr.	gr.	gr.	gr.	gr.	gr.	gr.
Urine	1330	302	109	11	34	38	110
Excrement	14250	3525	1364	180	1329	78	575
Total	15580	3827	1473	191	1363	116	685
Total of fodder	25770	8392	3938	446	3209	139	672
Difference	10190	4565	2465	255	1846	23	13
Sense	–	–	–	–	–	–	+

WATER RECEIVED BY THE HORSE in 24 hours	kg.	WATER VOIDED BY THE HORSE in 24 hours	kg.
From hay	1.035	Urine	1.028
From oats	0.448	Excrement	10.725
Drunk directly	16.000		
		Water voided	11.753
Water received	17.483	Water received	17.483
Water voided by pulmonary and cutaneous perspiration			5.730

1846–47 crisis: "There is a natural right to eat when one is hungry; the government should lower the price of wheat and raise weavers' wages, or wagon drivers will be killed and farms will be burned."[72] The workers' journal *L'Atelier* again tied this right to higher wages: "Under these circumstances, we say that a natural consequence of a rising [price of] subsistence is a raise in wages. . . . This winter, wage agitation will be fed by the scarcity of subsistence."[73] Following the 1848 revolution, the provisional government echoed these words when it "committed to providing for the worker's existence through work."[74] Subsistence and wages were tightly linked in workers' reactions to the midcentury crises. Traditional charity and paternalistic measures proved incapable of meeting this challenge.

Boussingault and Dumas offered a model for managing scarce resources: the farm. Pigs, cows, horses, and sheep served as experimental models for the measurement of humans' daily needs. Dumas and Boussingault employed the scale balance to measure flows of matter between bodies and their environment. As economic actors they aimed to minimize waste and unnecessary expense. They sought to identify the most economical food supply that maintained their animals' body weight.

BREAD AND MEAT

The two essential elements of animal life, in Dumas and Boussingault's view, corresponded to the two basic foods in a typical French diet. Bread replenished the carbon produced by animal combustion and released by respiration. Meat supplied the nitrogen excreted in the urine. To determine an ideal ration of bread they calculated the amount of carbon exhaled by a healthy adult. To set rations for meat they measured the nitrogen in the urine of men at rest and at work. Dumas conceived of a precise equation between men's minimum daily requirements and the body's excretions of those two elements. "By the means of experimental data, it is easy to say what is the minimum food appropriate for a man, and what kind of food he needs: knowing that he must burn carbon . . . and ammonia, on the one hand, and having determined the [chemical] nature of different foods, on the other hand, all one needs is a simple equation."[75] In this way, Dumas translated his abstract model of the chemical balance directly into human dietary standards. Since workers in the 1840s spent two-thirds to three-quarters of their wages on food, Dumas and Boussingault must have been aware that their measures also represented workers' basic needs.

To set dietary standards one needed to measure the carbon and nitrogen excreted through respiration and urine. Dumas and Boussingault's animal chemistry focused therefore on two types of experiment: the capture and analysis of excrement and of exhaled air. Dumas and Boussingault accorded "the highest importance" to measuring air, "upon which depends the most serious questions of nutrition, and thus of public economy."[76] Dumas undertook a series of self-experiments in which he filled a balloon with his own exhalations. After half an hour of breathing he stopped up the balloon, measured the volume of gas inside and weighed the exhaled carbonic acid in caustic potassium. He calculated that he expired three hundred grams of carbon per day.[77] Two of Dumas's protégés carried out a more elaborate procedure involving a sealed breathing mask on men and women of varying ages and states of health.[78] The three-hundred-gram figure became the basis for standard measures of dietary rations of carbon-rich foods, notably bread. Popular dietary guides calculated the amount of bread (approximately one thousand grams) required to fill that ration, thus supplying the other side of Dumas's "simple equation."[79]

Dumas used the same inductive procedure to determine a minimum amount of air that men must take in to supply the oxygen necessary for bodily combustion. "Numerous experiments have shown . . . that a man needs 6 to 10 cubic meters of fresh air each hour."[80] His students analyzed air samples from the streets of Paris, from bedrooms, hospital rooms, schoolrooms, military stables, a greenhouse, and the amphitheater at the Sorbonne after one of Dumas's lectures.[81] Boussingault repeated similar exhalation experiments on an adult man, a horse, a milk cow, a pig, and a sheep. He claimed that his measures should "permit us to set a limit for the capacity of dwellings and stables."[82] In each of these cases Dumas and his colleagues sought to determine whether the men and animals in these sites received an adequate "ration of air" to guarantee proper nutrition.[83]

The nitrogen in urine proved considerably easier to capture. Chemist Louis René Lecanu supplied the necessary figures in his analysis of the urine emitted by a sample of adult men and women, children, and the elderly.[84] Dumas and Boussingault referred to Lecanu's adult male subject and determined a minimum daily ration of fifteen grams of nitrogen.[85] To ensure that he did not miss any nitrogen that might "escape" through respiration, Boussingault drew from his experience with plant chemistry. He constructed glass-domed apparatuses that allowed him to capture the entirety of an animal's air and excrement. Boussingault enclosed turtledoves in a glass jar connected to a balloon that collected its exhalations and a removable screen by which its excrements were collected and its feed weighed and rationed.[86] This experimental setup became the basis for chemist Henri Regnault's elaborate large-scale containers for large animals and, later, for German chemists Carl Voit and Max

TABLE 2 Jean Baptiste Dumas, *Traité de chimie appliquée aux arts*, 542

	URINE EXCRETED IN 24H (GRAMS)	CARBON BURNED IN 1 HOUR (GRAMS)
Children 3 years old	13.5	5
Men	28.1	11
Women	19.1	6.5
Elderly	8.1	7.4

Pettenkofer's human respiratory chamber.[87] Nitrogen use, they concluded, increased with activity; thus they confirmed the popular notion that heavy laborers required more meat.[88]

For those lacking the equipment to analyze the chemical composition of foods, Boussingault suggested, body weight could serve as an indicator of dietary balance. Following Lavoisier, Boussingault observed that "an animal on a maintenance ration return[s] every day to the same weight it had the previous day."[89] A man who ate with regularity would remain consistent if weighed at the same time each day.[90] Boussingault undertook such weighing experiments on his horses in order to establish this principle in spite of apparent "accidental variation" and "irregularities" which appeared from one day to the next.[91]

In these procedures, Dumas and Boussingault drew from a long tradition of agronomic weighing and measurement. Agronomists, following Lavoisier, were among the first to recognize the utility of the scale balance in regulating diet. A scale balance registers the body's inputs and outputs, the exchanges between the body and its environment. Weight measures can be compared synchronically and diachronically, around a norm, across different groups or over time. The scale records body weight before and after a meal or at the beginning and end of a given period. In chemistry as in medicine, the scale balance was employed to track movements and transformations that could otherwise not be seen. Digestive processes and organs remained something of a mystery in the mid-nineteenth century. The scale allowed one to bracket that movement and draw up an account of inputs and outputs.

Dumas and his colleagues measured the carbon and nitrogen released by their experimental subjects. On the basis of their theory of the balance of matter they converted these "natural" measures into concrete dietary guidelines. They compared their theoretical rations with the actual rations in the army and schools. Dumas noted with satisfaction that standard military rations (including 285 grams of meat and 750 grams of bread) corresponded closely to his calculation of a man's minimum daily requirements of three hundred grams of carbon and fifteen grams of nitrogen.[92] Conversely, they used their scientific results to argue publicly for changes in dietary policy. Dumas and his protégés participated in state commissions charged with reviewing the rations of marines, soldiers, hospital patients, and schoolchildren.[93]

"Bread and meat in sufficient quantity, pure air and pure water: these are the foods that maintain the health of the individual and thereby improve the human race." Dumas did not hesitate to use his scientific results as warrants for pointed social commentary. "From these foods the human race obtains a condition of well-being that results from the appropriate balance between our physical and moral forces."[94] If a nation failed to supply its population with the necessary quantities of carbon/bread and nitrogen/meat, "the conditions of public hygiene are altered . . . [and] this state of suffering . . . sadly is often the case."[95]

RECYCLING AND REPRODUCTION

Animal ingestions and excretions were but one station in a globally circulating economy of matter. Dumas and Boussingault articulated a conservationist view of matter. A fixed quantity of matter moved between the environment, plants, and animals. The atmosphere transported the essential chemical elements of life to the soil. Plants fixed and organized these elements. Plants, for Dumas and Boussingault, represented "reduction apparatuses . . . [that] seize . . . carbon, hydrogen, nitrogen, ammonia" and convert that atmospheric material into nourishment.[96] Plants, living laboratories, carried out the work of food production; they "fashion organic or organizable material that they give to animals."[97] Animals, in turn, act as "combustion apparatuses," whose by-products "carbonic acid, water, ammonium oxide and nitrogenous acid" return to the air to begin the cycle once again. "Thus closes the mysterious circle of organic life on the surface of the globe."[98]

This vision of combustion and reduction, animals and plants, was not new. Emma Spary shows us that circulation and recycling played a key role in eighteenth-century theories of nutrition. George-Louis Leclerc de Buffon proposed in 1753 that organic matter traveled in an eternal cycle through plants to animals to humans, "from body to body."[99] Dumas and Boussingault applied that model to the inner economy of living beings.[100] Animals were passageways in Dumas and Boussingault's *circulus*. Plants did most of the work of "organization," transforming atmospheric chemical elements into usable nutrients. Animals ingested and incorporated those nutrients into their tissues; Dumas and Boussingault limited animals' action to that of combustion.

This view formed a point of contention in one of the greatest scientific controversies of the day, their debate with German Justus von Liebig about animal fat.[101] Liebig and his allies considered animals as "veritable laboratories where fat is produced."[102] Dumas denied that animals could "modify or transform" the nutrients that they ingested. Instead he submitted that nutrition should be understood primarily as the "simple passage of nutritive matter."[103] Even when bees were shown to produce wax on a diet of sugar, and geese to fatten on a diet of corn, Dumas insisted that these were exceptional instances.[104] Animals oxidized their food by combustion and assimilated other nutrients into their bodily structure.

Dumas and Boussingault maintained their vision of a continuous circulation of matter. "All the changes that operate on the surface of the globe are due to combinations that form and break. The matter in a prairie's green carpet . . . becomes part of the animals that it feeds; a few days later, perhaps it will pass into our own organization, from which it will enter the atmosphere and from there be rendered to plants, who will reproduce a new vegetation."[105] Nutritional needs could thus be measured on the basis of the transit of matter into and out of the body.

Some doctors and physiologists objected to the chemists' practice of bracketing differences between individuals and ignoring the internal function of bodily organs. Vitalist doctors refused to view nutrition as a question of "addition" and subtraction. Following Pierre Cabanis, they argued that nutrients operate to "excite certain functions in the organs," such as intelligence, energy, or impulsion. Physiologist François Magendie fed his experimental subjects (dogs) a diet of only sugar or oil, which contained one or two essential principles, and noted that the animal subjects fared badly. A single nutritive principle could not maintain life for long, he argued; "what we know of its chemical composition does not suffice to allow us to predict its nutritive value."[106] Claude Bernard, Magendie's successor, developed his research program on the physiology of digestion partly out of frustration with Dumas's methods.[107]

For Dumas and Boussingault, animals'—and people's—needs were defined by their place in the *circulus* of matter. Two types of animal product were essential to the cycle of life: exhalations and urine. Urine carried ammonia back into the atmosphere, ready to serve once again as the building block for plant life. Exhalations released carbonic acid, the by-product of animal combustion, into the air. "Man is a marvelous machine . . . for he rejects back into the general [natural] economy the products which serve to reconstitute the fuel that he has consumed." Through the intermediary of plants, all of the nitrogen and water released by animals' urine is converted back into "precisely the amount of carbon that they have consumed."[108] Dumas concluded that the task of agriculture is to "remake, with men's urine, and by means of carbonic acid in the air, the wheat that man eats."[109]

Reproduction, in Dumas's *circulus*, took place primarily through the organization of chemical matter in plants. Men "replaced" their bodily substance with food. Dumas referred his dietary standards unreflexively to a standard adult man. He implied that by ejecting chemicals into the atmosphere, a man could in a certain sense reproduce himself.[110] Man's urine releases ammonia into the air and soil, matter which enters the wheat that provides later for his daily subsistence. Eating and reproduction involved the same processes of loss and replacement of matter. The material of life appeared to perpetuate itself sexlessly. Agronomic measures of need effaced sexual difference; women and children barely registered.

The agronomic body was a scientific model for how men alone "reproduce" and create value in the cycle of nature. It served as a physiological analogy to the social-economic theory of wages. The parameters of human needs, in this account, were set not by social institutions or by the free market but by the natural flow of matter. Dumas and Boussingault saw the cycle of matter as a great productive machine, which men must shape and direct to serve human ends. Scientific expertise would act in the service of the state and of social stability. This was a romantic-bureaucratic model of life, economy, and science.

CHAPTER 3

Social Reform

Social reformers of the July Monarchy (1830–48) and Second Empire (1851–70) turned to science to resolve problems of population, wages, and the reproduction of labor. How to ensure a basic level of subsistence in a free market? How to manage a labor economy that produced migrations, waste, and inequalities, and that seemed to corrupt or damage workers' ability to replenish themselves? How to establish a stable, industrious, efficient social order in an unstable economy?

Chemists like Dumas and Boussingault promoted dietary standards as guidelines for managing labor and poverty. They were acutely aware of their social mission to help "feed the poorer classes" and explicitly stated the problem of nutrition as an issue of class.[1] They applied their chemical-agronomic science to the two great social problems of the day: the social role of the state and the wages of labor.

This part of my story is about an emerging techno-science of the welfare state. Dumas and Boussingault's experiments contributed to a set of techniques for managing population and labor: use of the scale balance, input-output equations, food rationing, and the science of work. Scientists began to trace relationships between consumption and labor, home and work, production and reproduction. In this early phase, around the mid-nineteenth century, institutionalized populations such as schoolchildren, soldiers, and prisoners served as model subjects. By the end of the nineteenth century, these techniques spread to a far wider population.

Poverty, in the mid-nineteenth century, was widely understood as a disease. The field of public hygiene bound moral values, physical geography, and poverty to bodily health.[2] Medical statisticians collected data on mortality and morbidity by age, sex, neighborhood, climate, and a multitude of characteristics. In this process they created new social categories and new modes of social intervention.[3] Hygienists aggregated population data along selected traits. They described the effects of social inequality on health and life expectancy, thereby medicalizing poverty and class.[4]

Hygienic reformer Louis René Villermé famously used statistical methods to link public health and social inequality. His statistics treated the poor as a collective entity, with a collective fate.[5] In the 1820s and 1830s Villermé collected data on rates of sickness and death in regions and cities across France. Within Paris he mapped population data by neighborhoods and individual streets. He attempted to isolate the cause of different mortality rates by dividing populations according to a number of characteristics: climate, age, sex, urban or rural environment. Discounting all of these elements as primary causes, Villermé concluded that "wealth, affluence and poverty are, for the inhabitants of different arrondissements in Paris, the principal causes (we would not say the only causes) that we can observe in [rates of] mortality."[6] He established similar correlations between poverty and sickness, physical deformity and inferior height.[7] Poverty, thus, was a cause of illness and death.[8]

Villermé's work on mortality implied that the poorest class of the population did not have the means necessary to live. However, he rejected the conclusion that workers' wages were to blame. Instead he suggested that workers themselves needed to adjust their lifestyle and behavior. In his later work he turned away from aggregate statistics to undertake close ethnographic observation of urban textile workers. He collected local data on wages and the cost of living in several textile areas. He allowed that the textile industry's lowest wages were just barely adequate to provide for a working family, if it were "industrious and of good conduct."[9] The more workers earned, he warned, the more likely they would waste their incomes to satisfy desires and "debauchery."[10] He worried that urban industrial neighborhoods propagated vice and consumer desire. Villermé argued for a salutary return to the land and rural industry, where workers lived abstemious and healthy lives.

Hygienic reformers like Villermé assembled an impressive array of statistical indicators, which led them to a conundrum. They had created a statistical entity, the "poorer classes." Such an aggregate view of poverty implied that it was a collective, social problem. A social illness required a social cure. However the prospect of government intervention offended many hygienists' liberal politics. Beginning in the 1840s hygienists sidelined measures of wages and needs, with the exception of women's and children's labor (which they considered a drag on family subsistence).[11] Instead their statistics focused on families, child rearing, domestic and public spaces, consumer habits and moral behavior.[12] Over the course of the nineteenth century, the economic elements of hygienists' studies gave way to environmental issues, contagion, and bacteriology.[13] The statistical approach to poverty appeared a dead end.

Historians of public hygiene tend to portray this turn as a scientific closure of the wage question. Collective statistics implied that wage differences were social and had deadly consequences. But hygienists' free-market orientation led them to refuse any state intervention in the economy.

But chemists and agronomists offered an alternative science of wages. Instead of concentrating on statistical aggregates, chemists weighed and analyzed individual bodies. They employed a transactional model. Gerard Jorland suggests that historians of nineteenth-century French hygiene, mainly concerned with medical statistics, underestimate chemists' importance to the history of social reform.[14] This chapter provides evidence for his claim. Dumas and Boussingault's chemical model had a significant impact on social policy across the July Monarchy and Second Empire.

Lavoisier's dilemma—why workers expended more of themselves but consumed less food than the wealthy—now appeared to threaten France's very existence. Where the laboring classes worked for wages, workers could not directly supply their own food, clothing, and housing. Unlike tenant farmers or artisan-peasants, workers who owned no property or capital had no independent means to provide for their own subsistence. They had to purchase the necessary goods on the market. Yet economists taught that wage rates were set on a free market, as a function of supply and demand. There was no guarantee that wages would cover the needs of a working family. French social reformers worried that the laboring classes would not be able to reproduce themselves. How could the nation reconcile liberal democracy and the ideals of the French Revolution with a free market economy? Under a regime of the free wage contract, workers risked being "enslaved to their needs."[15] The French Revolution promised equality; the wage economy created patent inequality.[16] Many feared that contradiction would lead directly to another revolution.

SCALE BALANCES

In 1854, a scientific commission made surprise inspections in school canteens across France, armed with a scale balance. "A few minutes before teachers entered the canteen," it reported, "without having made our intentions known . . . we asked that the pieces [of meat] destined for a table for students

FIGURE 3. Roberval's Balance. © Jacques Boyer/ Roger-Viollet/ The Image Works.

of the *petit collège* be placed on the balance."[17] The commission then measured pieces destined for teachers, presumably to ensure that they did not skim from their students' lunches. Inspectors seemed relatively indifferent to the students' general health or satisfaction with their meal. Instead they trusted that the scale balance, thrust upon the unsuspecting cooks and teachers, would reveal all that they needed to know about diet.

The 1854 commission's reporter, Dr. Pierre Honoré Bérard, justified the use of the balance as a scientific and pragmatic tool. The scale balance, he argued, brought "speculative [scientific] ideas and [general] practice" into "agreement."[18] Bérard taught at the Paris medical faculty, where his course on physiology focused on problems of nutrition, digestion, and the Dumas-Liebig debates.[19] Bérard's lectures show Lavoisier's influence on nineteenth-century medicine. Environmentalist theories of disease dealt with the action of external agents upon the human body. Medical students learned that various external influences were "modifiers of health"; diet appeared as the most important category of "ingesta."[20] Lavoisian chemistry offered a technology for measuring these effects. A food's "nutritive value," Bérard explained, depended on its reaction with the oxygen that a body breathes; that is, on the food's ability to fuel Lavoisier's combustion chamber.[21] Nutrition was an issue of "acute interest," Bérard told his students, for both science and public hygiene.[22]

Bérard began his lecture with a rhetorical question: "Can a single food nourish a person completely?" By this time the answer was well known, thanks to twenty years of debates over the nutritive value of gelatin. Early nineteenth-century manufacturers of gelatin claimed to have discovered a cheap substitute for meat, extracted from otherwise inedible animal parts. In 1831 a commission was appointed by the French Academy of Science, with a mandate to determine the usefulness of animal gelatin for feeding the poor. Gelatin, promoters claimed, could provide low-cost sustenance to state-supported populations, hospital patients, schoolchildren, and soldiers.

The Gelatin Commission of 1831 validated the use of chemistry in public hygiene and the management of poverty. Through the Gelatin Commission, chemical methods entered the social realm. The academy named three chemists, a physician, and a physiologist to test manufacturers' claims. The commission's work spanned nine years, as its mandate expanded to the general question of how to measure nutrition. Physiologist François Magendie undertook large-scale feeding experiments on dogs in the academy's basement. Chemists Michel Chevreul and Louis-Jacques Thénard analyzed gelatin's chemical components. Jean Baptiste Dumas, then Thénard's protégé, was the youngest member of the commission. These men recognized the public implications of their work, on "the definitive solution to a problem in the interest of all of humanity."[23]

Rather than arriving at a definitive solution, however, the commission's work opened up a set of questions that scientists and public officials would grapple with for decades. Three central issues emerged: how to analyze a food chemically, how to determine its nutritive value, and how to apply that scientific knowledge to the practical problem of feeding the poorer classes.[24] Dumas and his colleagues would refine these questions over the course of the following decades. State dietary policies, under the July Monarchy and Second Empire, closely followed their conclusions. The chemical scale balance penetrated the social realm.

Weight measures represented animal bodies as material objects, commensurable with other material things. The scale registered bodies as "samples of matter" devoid of individual natural histories, eccentricities, and conditions.[25] Until the late seventeenth century, the scale balance was used primarily to measure goods and substances, money and commodities. In eighteenth-century chemistry and physics, the scale balance symbolized stability. It was used to describe conservation and equilibrium, in a system where variations appeared as oscillations around a stable norm.

By the eighteenth century weighing became a novel public activity in some places; scale balances used for coffee were set up in London coffeehouses for patrons to measure and record their weights at intervals of days or years.[26] The practice of weight watching drew from the seventeenth-century static medicine of Sanctorius Sanctorius, who invented a hanging scale chair to measure bodily losses due to what he called "insensible perspiration."[27] But the scale balance did not find favor with French medical practitioners, who considered dietary qualities (such as succulence and fermentation) more important than quantities. "In Paris, weighing was a technique commanded by grocers and apothecaries, not physicians."[28]

In agriculture, by contrast, the scale balance served both medical (veterinary) and commercial purposes. Animals were weighed at the point of sale like wheat, potatoes, or any other commodity. As the science of agronomy spread across France, Britain, and Germany, weight measures were used more dynamically. Farmers began to apply principles of scientific rationalization to their livestock management. Their aim was to lower the cost of inputs—animal feed—and to increase the output of valuable products—milk, eggs, meat for slaughter, manure for fertilizer, and work. In England J. B. Lawes weighed animals' bodies and manure under varying diets in order to maximize fertilizer production.[29] German agronomist August Weckherlin weighed several cattle breeds under various dietary conditions with the aim of increasing milk output.[30] In the hands of agronomists the scale balance became an instrument for monitoring, regulating, and optimizing the animal diet.

The nineteenth century was the age of authority for the scale balance. The July Monarchy revolutionized the relationship between commerce and science. The July Monarchy, like the Second Empire, was run by "technological men": scientists, engineers and state-trained functionaries.[31] The majority of parliamentary representatives had experience in local government. They sidelined intermediary professional and charitable bodies and centralized economic and social administration.[32] They conceived of the state as a social power, acting in the interests of business, industry, and a meritocratic elite.[33] They used scientific methods to solve social questions.

Louis Philippe and Napoleon III both encouraged the development of positivist social science.[34] They believed that a strong and prosperous society required a rational, utilitarian state. "Whether concerned with customs, penitentiary or penal organization, higher education, cadastral or land policy, there was a constant concern to ground all public action in positive philosophy, reasoned argumentation, classification, measurement."[35] Commissions, surveys, and statistical studies abounded on all subjects related to public welfare and policy—schools, charity, prisons, urban food consumption, commerce, population.

The July Monarchy put the metric system into practice nationwide. Each measure had a natural scientific referent: a meter, a segment of the earth's meridian; a gram, the mass of a cubic centimeter of water at 4 degrees Centigrade. The metric system was born of the French Revolution, though initial attempts to impose the system nationwide resulted in popular rebellion. Only in 1840, when the July Monarchy outlawed all other measures, did it become standard. The metric system was intended to erase the innumerable local standards in use across France and thereby to facilitate the free and transparent flow of commerce.[36] Universal measures allowed equivalences between all sorts of goods and objects. Through the metric system and the scale balance, chemists and laymen could bring together "disparate substances in a single equation."[37] Popular measurements, such as weights of goods for sale, were comparable to scientific measures. Agriculture, commerce, and chemistry shared the same units of measure. State-mandated inspectors applied the scale balance to schoolchildren, prisoners, soldiers, and others under their purview. Commissions with titles such as the "Council to Monitor the Revision of Diets" sprang from hospital, prison, and school administrations.[38] Some designed nutritional standards; others toured the country in order to verify conformity. Decrees and directives specified food rations down to the last ingredient, as witnessed by a missive from the Parisian Public Aid Administration on permissible quantities of salt and seasoning.[39] The scale balance bridged chemical dietary science and social practice. It represented a concerted effort to translate chemical-agronomic guidelines into real meals.

Bérard, in his report on school meals, subscribed to a chemical vision. "Science teaches us that in the various actions that we call life, while an animal breathes, eats and maintains its temperature, it moves and feels, organic matter is destroyed." The chemical view of life propounded by Dumas found a real-world correlate in the measurement of schoolchildren's meal size. "The scientist," by whom Bérard clearly referred to Dumas and Boussingault, "collects, analyzes and weighs the products of this bodily decomposition . . . and from those observations it deduces the nature and quantity of foods necessary to repair those losses."[40] Bérard submitted that administrators ought to follow this balancing technique when preparing each school meal.

The 1854 commission forcefully recommended the introduction of the scale balance into every school canteen. The Ministry of Education followed

Bérard's suggestion that it "set the weight of cooked meat to be served to students of different schools." The ministry's directive demonstrates the level of precision to which state administrators designed their rations.

> Dinner consists of a main course of meat on "fat" days, and of eggs on lean days. One should add a second course of vegetables, conserves, salad, prunes or cheese; on lean days the second course should be composed of substantial vegetables, of grains. The amount of cooked meat, boned and trimmed, delivered to each student is regulated thusly. . . . For very big [children]: 80 g.; for big [children]: 70 g.; for middle-sized: 60 g. [etc.].[41]

The scale balance would guarantee the conformity of every meal. "The headmaster, or another delegated person," recommended Bérard, "will, as often as deemed necessary, ensure that the distribution [of meat] conforms with the regulations." Just as the commission had done, the headmaster should introduce himself unexpectedly into the canteen kitchen fifteen minutes before a meal and place a sample of meat servings upon his scale balance. Bérard admitted that the idea of using a scale balance to apportion school meals was not new, "but it seems not to have been executed at all rigorously."[42] By the 1860s that situation had reversed. Another commissioner reported in 1868 that school "menus everywhere are organized according to regulations on the quantities [of food] to be distributed to students. Everywhere, I saw scale balances operating."[43]

AIR RATIONS

On a cold winter day in 1843, a young chemist sat in a cell at Paris's Conciergerie prison with a pinwheel device in one hand and a container of excrement at his side. Endowed with an unusually sensitive nose, Félix Leblanc was called to this duty by a special commission composed of France's most prestigious chemists. As Leblanc sat breathing in isolation, Dumas (his thesis adviser), Boussingault, and two senior colleagues stood outside and varied his air supply.

A "perforated chair," designed to resemble a prisoner's latrine, stood at the center of the room. It contained a bucket of water and a sample of bodily waste presumably collected from some accommodating animal. A long tube led from the base of this makeshift potty into the room next door. Outside, several burning candles generated a draft, which pulled air down around the bowl of waste and out of the cell. On the opposite side of the cell, a flue led outside to a stove that heated the air.

These two holes in the wall offered Félix Leblanc's only contact with the external world. The cell served as an oversized bell jar, an instrument for regulating airflow. No breeze, chill, or noise disrupted the solitude of his chamber, whose door and windows had been tightly sealed. Four expert scientists monitored the evacuation tube and adjusted the size of its opening. They sniffed the exiting air, collected it in glass balloons, and measured its humidity and chemical composition.

As the experiment progressed, an overwhelming smell began to waft out of the perforated toilet bowl. Measuring instruments picked up rising levels of ammonia in the air exiting the cell. The poor young man, according to official reports, began to feel a considerable malaise. Yet the attending scientists hesitated to increase airflow to the cell. Six cubic meters of air delivered hourly seemed largely sufficient. Despite Leblanc's expressions of disgust, they held fast at that level.

In the third hour of this ordeal his adviser and fellow commissioners finally relented and increased the cell's air volume. When the airflow grew to ten cubic meters per hour, the record shows, Leblanc felt relief at last. He opened and closed the latrine cover in order to time the evacuation of putrid odor. It was reported with satisfaction that the ten-cubic-meter airflow whisked smells away in a mere twenty minutes. The "infectious odor" of air exiting the draft tube signaled that the cell had been washed of all noxious substances. (When the bucket of water was taken away, however, and the excrement stood alone in the latrine, no amount of air could remove the stench.) Once the airflow had increased to ten cubic meters, Leblanc passed the following seven hours comfortably in his cell—or so the commission's secretary claimed.[44]

To what end did an intrepid young chemist find himself measuring air in a stinky cell? The results of Leblanc's ten-hour ordeal, and several more trials like it, were used to establish a scientific minimum volume of air for each inmate in France's newest cellular prison.[45] The scientists who were called to the Conciergerie in 1843 by the French interior minister were presided over by the astronomer-politician François Arago and included several of the most famous scientists of the day.[46] Their mission was to establish a standard for airflow in closed rooms, and to select a ventilation system for the new Mazas prison.

Thanks to Leblanc's discerning nose, the commission fixed a "ration of air" for each individual prisoner: "The preceding experiments appear to demonstrate . . . that a ventilation of less than ten cubic meters per cell and per hour would be insufficient to maintain the air in an appropriate state of purity . . . and above all to counter the diffusion of disagreeable smells that emanate from the seat which stores the prisoners' excretions. . . . Therefore we believe that this figure [ten cubic meters] . . . should be considered a lowest limit for the conditions in which a prisoner should be placed."[47]

The design of the prison cell experiments that Boussingault and Dumas performed for the Mazas Commission in 1843 derived from their scientific work. The precise accounting of chemical input and output in plants and animals required strict environmental controls. Dumas praised Boussingault to the academy for his well-regulated experiments: "When [he] wishes to identify the influence of water or air on a plant, he puts it in a closed jar in contact with these two purified substances, and he analyzes the elementary [chemical] composition of the plant before and after it is introduced into this apparatus, which conceals it from all foreign influence."[48]

Indeed, Boussingault's apparatus—a bell jar with input and output tubes—provided a template for many subsequent studies. Each of these experiments sought to establish standard "rations" of food and air. On several occasions Boussingault "sequestered" his pigs and cows in individual stalls or "cells" in order to measure a minimum animal diet.[49] He and chemist Henri Regnault enclosed birds and dogs in glass bell jars so as to record the amounts of carbon and oxygen that they inhaled and exhaled every hour.[50] In 1843 Félix Leblanc found himself in such an experimental setup, in the position of the bird.

Eugène Péclet, a scientific innovator in indoor heating and a member of the prison commission, established a similar standard air volume for schools. Péclet forced a class of 180 schoolchildren, ages seven to ten years old, to spend a summer afternoon with all their doors and windows sealed. By the time the class teacher begged Péclet to open the windows, he had determined a minimum sufficient airflow: six cubic meters per hour.[51] A few years later, the Ministry of Education sent another hygienic inspector out to examine schools all over France. He tested Péclet's figure, updated it (allotting a more generous sixteen cubic meters to each student) and advocated the placement of pinwheel air counters in every classroom.[52]

"Science has determined the precise quantity of cubic meters of atmospheric air that each individual needs for respiration to take place without any trouble. . . . [Yet] this law has rarely been followed by architects," complained a *Treatise on the Sanitation of Large Cities*.[53] The authors offered a rule for housing design: the minimum size of each room should equal the number of its inhabitants, multiplied by six cubic meters of air volume. Science had responded to the call for hygienic living standards; it now fell to architects to carry out these "laws."

In 1848, founding year of the Second Republic, private housing became a full-fledged public political issue. Parisian landlords briefly became targets of the rebellious popular spirit that had ousted the July Monarchy. In some neighborhoods, bands of residents heckled landlords into letting rent payments slide.[54] Although the forces of property and order soon rectified that trend, the newly elected bourgeois parliament felt compelled to intervene. London had recently promulgated rules for the organization and minimum volume of private apartments, and some called for Paris to follow the English example.

> While science demands twelve to fourteen cubic meters of air per person, there are renters in some buildings who spend their lives in dwellings that barely provide them with three or four cubic meters to breathe. We say that such dwellings should be prohibited. A GUILTY INDUSTRY speculates on the primary good that God gave to man, the air that he breathes, air without which life could not continue![55]

The prefect of Paris responded, predictably, by calling for a commission. The Paris Council on Public Health obliged his request with an "Instruction on the Means to Insure the Sanitation of Dwellings." Council members warned that city sanitation depended in large part on the purity of the air. "It is in dwellings with unhealthy air," they wrote, "that epidemics are born and spread with greatest intensity, which then wreak havoc on entire cities."[56] On the council's recommendation, the prefect set norms for *garnis,* cheap furnished rental apartments destined for the transient working class: "It is important that the number of beds placed in each bedroom shall be in proportion to the room's dimensions, such that there is at least fourteen cubic meters per person, independent of ventilation."[57]

The Council on Public Health did not provide any scientific reasoning for this new measure. It seems clear, however, that the prison air experiments five years earlier must have influenced this outcome. Indeed the author of the council's recommendation, Dr. Alphonse Guérard, soon would head a second Mazas commission.[58] As a doctor at the Hôtel Dieu hospital, he would have been aware of ventilation experiments carried out there in the eighteenth century. In Guérard's report, Dumas and Boussingault's findings took on a new dimension and new meaning. The chemists' ten-meter measure became a standard fourteen cubic meters of living room per person.

Despite public clamor and health officials' intentions, the prefect's missive remained more theoretical than legal. This ambitious attempt to transform workers' living conditions had an even shorter life than the Mazas experiment: the decree was never enforced. A year later, a follow-up municipal commission threw up its hands: "Is it possible to establish precise rules for dwellings and their construction, to determine the width and dimensions of courtyards, alleys, lodges, bedrooms etc.? We can only call this point to the attention of architects and investors, which we have done. . . . In sum, the commission considered that this question stood outside its mission as defined by the law."[59]

A national law in 1850 required inspection of "unsanitary dwellings." This decree elided the whole question of quantitative standards and simply condemned all forms of *insalubrité*. The inspectors appointed to enforce the law rarely entered private buildings; they contented themselves with assessing healthful or dangerous appearances from the outside.[60]

Regulations on air supply thus largely applied to state institutions: the prison and school. Under the July Monarchy state officials became increasingly concerned about corruption and contagion in prisons' common spaces. Large workshops and dormitory barracks risked propagating infectious and immoral contact between prisoners. French administrators worried that their barrack-style prisons were retrograde in comparison to Bentham's Panopticon and new American designs.[61] "Cellular imprisonment," the Interior Ministry declared, "is the most effective remedy against the excess of corruption generated by the current state of [France's] prisons."[62] From Limoges, a prison director wrote that "the establishment of cells would be a real improvement; it would guarantee the [prisoners'] morals, health, even their lives."[63] The cell offered protection from outside influences and a perfectly regulated environment.

Having decided to build an isolated space for each inmate, officials focused on optimizing its parameters. "Each cell," ordered the 1844 law on prisons, "should be sufficiently spacious, healthy, and aerated."[64] In the context of the prison, "sufficient" signaled the lowest necessary ration for hygiene and survival. The heavy budgetary demands of the new prison design could only be accommodated if each cell was reduced to its smallest hygienic volume.

The Mazas prison, which replaced an aging detention center in Paris's Faubourg Saint Antoine, was one of the first to implement the government's new prescriptions on cellular design.[65] The prefect of Paris called upon Dumas, Boussingault, Leblanc, and their colleagues on the commission to set a scientific standard for the prison's air supply. By trial and error, as we have seen, they settled on an hourly figure of ten cubic meters of air per person. This standard determined the government's choice of building and ventilation plan.

In May 1850, the Mazas panopticon opened its doors. Architects Emile Gilbert and Jean François Lacointe, champions of utilitarian design, built six prison blocks radiating from a central surveillance tower. Each of the individual cells contained a ventilation tube that pulled air from a flue in the cell's ceiling, down through the prisoner's latrine and into storage barrels in the building's basement. The system's air supply was set and tested by the commission.

To prepare the establishment to receive its new inhabitants, the prefect of Paris spared no caution. Fifty indigent citizens were "borrowed" from the poorhouse at Saint-Dénis and assigned the (seemingly involuntary) role of prisoners for a night. Two members of the Mazas commission equally locked themselves up for a few hours in order to test the prison's habitability and the validity of their conclusions.[66] They gave the all-clear. Hundreds of "preventive prisoners," men awaiting trial, were transported on site, "washed and disinfected as necessary," and led each to his own cell.[67]

Within a month, things went wrong. The left-leaning newspaper *Le Siècle* recorded widespread rumblings among the inmates, who were mainly political detainees. Prisoners found the cellular system unbearable. In the heat of the summer air, many tried to open their windows. Prison administrators found this unthinkable, and the windows remained sealed shut. Nothing could countenance unapproved changes in the building's scientific ventilation.

Journalists at *Le Siècle* denounced the Mazas prison's "barbarian punishment of total isolation" and the "tortures of . . . lacking air and space."[68] The newspaper claimed that three "victims" had already succumbed to the prison's bad air. In the face of public scandal, a new commission was called and new experiments were ordered. Scientists again brought pinwheel counters into the cells and measured the airflow to ensure its adequacy. A group of commissioners spent an afternoon smoking cigars and pipes in one cell to observe the speed of smoke removal. A third group of experts conducted interviews with prisoners.

Predictably, all commissioners reported that the system functioned perfectly. They asserted that certain prisoners' discomfort and disaffection derived entirely from the fact that they had used their latrine wrongly or blocked the ventilation hole. Some inmates, they claimed, were naturally too stinky for any ventilation to remedy. On the whole, they found, prisoners enjoyed the new facilities. Several were quoted to declare, in an excess of enthusiasm, "they would consider themselves quite lucky if when they were free they could always be assured of such a comfortable dwelling."[69]

Still, the rumblings did not cease. Within two years, prison administrators caved in. They ordered the manufacture of 1,225 wooden plugs with which prisoners could block the ventilation tubes in their latrines and, finally, open their windows.[70] The commission's ten-meter standard succumbed to the resistance of its human subjects.

MAINTENANCE RATIONS

In May 1839, the French interior minister, Adrien de Gasparin, published a decree on prison diet. An agronomist and a politician, Gasparin understood the stakes involved in creating scientific measures of dietary need. When he served as prefect of Lyons in 1834 he had marching orders to crush worker agitation over wages. One year after his tenure as prefect, Gasparin had entered national government; he was named minister of commerce, agriculture, and industry and, three years later, minister of the interior. Like Boussingault, also present in Lyons in the early 1830s, Gasparin witnessed firsthand the surge of popular demands for a right to subsistence. He spent a good part of the following decades developing a model of nutritional measurement that might invalidate such claims.

Under Gasparin's leadership the Interior Ministry ordered prison canteens to reduce their supplies and to provide inmates only with rationed quantities of bread, boiled potatoes, butter, and cheese. Wine, beer, cider, and tobacco were absolutely prohibited. Gasparin's policy employed diet as a form of punishment: access to rations and supplementary food depended on an inmate's prison sentence and his good behavior.[71]

Prison meals drew the attention of administrators, philanthropists, and doctors and polarized opinions on nutrition and state administration. Prison doctors published a proliferation of articles on nutrition in journals such as the *Annals of Public Hygiene* and the *Bulletin of the General Society of Prisons*. Alexandre Hurel, doctor at the Gaillon penitentiary, contended that prison rations should be increased. He drew support from "remarkable" research in the field of nutrition, wherein "science on one hand, and daily experience on the other, give a solution which one should consider sufficiently accurate."[72] Hurel described diet in terms drawn from Boussingault and Gasparin: "For us, the best means [to determine prison diet] . . . is to take the maintenance ration as the basis for nutrition. In this way we will obtain the necessary limit between insufficiency and abuse."[73] In the 1840s the scale balance entered the prison.

A Genevan prison doctor weighed his charges biannually and published his results in the *Annals of Public Hygiene* along with a description of the prison's daily diet. (The latter consisted of a very large chunk of bread, vegetable soup, bouillon, and all the potatoes one could eat.)[74]

Many social reformers did not approve of Gasparin's punitive prison rations. Medical statisticians collected data to prove that prison mortality rates increased in the interval after Gasparin's 1839 decree. New instructions given to prison inspectors in 1843–44 suggested improvements in daily diet.[75] Gasparin's successor at the Interior Ministry, Duchâtel, declared in his "Instruction" on prison rations, "I want . . . the inmate's health to be looked after, that he receive all necessary care, that none of them . . . can complain of not having sufficient food."[76] The same minister took a tougher tone in a circular to the prefects in charge of supplying food: "If, in order to preserve health, it is necessary to give [prisoners] more abundant food, . . . you understand nevertheless, Monsieur Prefect, that there are limits that we cannot exceed without attracting justified criticism, as we would give offense to public morals. . . . In the study that I will undertake of . . . diet, I will be guided only by the intention to only allow what appears absolutely necessary."[77]

Despite this challenge from the field of medical statistics, chemical measures of nutrition retained their influence in state institutions. Reference works on public hygiene popularized chemists' nutritional measures. The entry on "Subsistence" in the *Dictionary of Public Hygiene* refers to chemical measures of rations alongside statistics of food consumption by city and region.[78] The 1854 *Elementary Treatise on Private and Public Hygiene* extensively cites both Justus von Liebig and Boussingault.[79] Popular hygienic digests such as these brought nutritional standards into the public realm. These works were directed at social reformers interested in the lot of poor and institutionalized populations.

Anselme Payen, a chemist and industrial entrepreneur, directed his study of nutrition to the attention of "all administrations of good works and of public aid":[80] "Governments, with a constant solicitude, follow the progress [of nutrition], because it leads to the achievement of the principal conditions of public hygiene; it tends to increase average life span, while gradually raising the level of well-being and the strength of populations. All charitable and public health organizations will find these scientific data useful."[81] Payen emphasized that nutritional standards each should correspond to a particular social group: "Men destined for hard manual work, like those who devote themselves to sedentary study . . . all have an interest of the highest order in following the positive rules of hygiene properly appropriate to each of their social positions; these precepts are all grounded in the same sure scientific bases."[82]

Payen confirmed that the minimum dietary standards set by Dumas closely resembled the actual amounts of nitrogen and carbon that he observed in prison and convent meals.[83]

*

Adrien de Gasparin did more than restrict prison diets. He applied agronomic science in order to standardize workers' wages. When Gasparin's political fortunes plummeted in the final years of the July Monarchy, he retired to the countryside and set out to summarize the state of agronomic science. The result, his *Cours d'agriculture*, became "the Bible of French agronomists."[84] Gasparin's contribution was not experimental but practical and political. Where Dumas and Boussingault referred to cost efficiency in general terms, Gasparin provided financial accounting for various animal and human diets. He turned Boussingault's animal studies into a measure of human life and labor.

As prefect in 1834, Gasparin had refused to intervene in the "free" wage contract between silk weavers and merchants. By the time he wrote his agronomic manual thirteen years later, he had shifted his view. Science, he believed, could establish a rational wage scale. Wages should correspond to the cost of reproducing labor.[85] His ideas on nutrition shaped public health and wage debates for generations.[86]

Gasparin posited two types of existence: maintenance and work. The maintenance body represented a pure consumer. "The maintenance ration," he wrote, is the amount "that would suffice to maintain a man's life without working and without losing weight."[87] In his paraphrase of Boussingault's maintenance ration, Gasparin brought the agronomic model to bear on people. "Maintenance rations" described an adequate diet for nonproductive populations and formed a baseline for workers' nutritional requirements.

The working body, on the other hand, expended its matter in labor. Its losses must be replenished at the end of each day. "Work rations" supplemented maintenance rations in proportion to a given quantity of labor. In addition to replacing a worker's "daily losses," Gasparin thought it necessary to "ensure the perpetuity of his work," by "providing for a replacement, that is to say, by maintaining his children."[88] These two kinds of ration corresponded to the two social problems that the agronomists sought to solve with nutrition science: population and labor.

Gasparin understood work as the loss of a laborer's physical matter. This definition drew from physicists' debates on the nature of work. Beginning with Lazare Carnot and Charles Coulomb at the end of the eighteenth century, physicists and mathematicians pursued a science of mechanical work. The framework for their research was economic: "how to measure the production and expenditure of men and machines, and how to optimize their use?"[89] Coulomb measured the "frictions" produced by various machines, including men turning a crank, and compared their relative "quantities of action."[90] Physicist Charles Coriolis and mathematician Charles Dupin sought to define a standard value for work and used human labor as their referent. Dupin, in the 1820s, defined an "adult-male equivalent," equal to the strength of an adult man, as a standard for all kinds of work. He used it to compare the relative power and efficiency of animal, machine, water, wind, and steam.[91] Even before thermodynamics provided a common unit for the work of animals and machines (energy), scientists understood animal and mechanical work in the same manner. They conceived of work in terms of physical expenditure.[92]

Gasparin referred to this model in formulating his ideas on nutrition as bodily loss and replacement. He included a table of measures of work performed by men, animals, and machines, alongside an account of their relative cost. Lacking a standard unit of human work (like the thermodynamic notion of energy), he set out to find one. "What," he asked, "could be a measure of the work that a man can perform?" He determined that the best unit of measure for a man's work was a cubic meter of lightly packed earth "dug up without effort, with the end of a hoe or with a shovel, and thrown aside at a height not exceeding 1.6 meters, which is the height of a cart (if one is filling a cart) or that of the edge of a hole (if one is shoveling a hole)." Based on "numerous observations of public works and day laborers," he calculated that "the average French man digs 15.6 cubic meters of such earth."[93] He called on scientists to provide comparable measures of man's muscular force in various countries and regions.

Gasparin measured the quantity of flour necessary to fulfill one of his standard units of work. "The maintenance of fitness necessary to produce one cubic meter of shoveled earth is . . . 45 grams of wheat, plus a fraction of the maintenance ration."[94] He compared his unit of measure of work with statistics of average flour consumed in the Loiret and the south of France. Using this formula he calculated the rations of flour that should be allotted to workers in the two regions. Northern workers dug more on average than their southern counterparts and thus deserved a slightly higher ration:

For work in the South: 0.67/15.31 wheat + 0.045k wheat = 0.887k
For work in the Loiret: 0.67/13.31 wheat + 0.045k wheat = 0.9k[95]

Gasparin described man as a very particular kind of engine. Animals were subject to friction and use, like all motors, and expended their very bodily substance in work. But unlike other motors, as Dumas wrote, man "ejects into the general economy the very products that serve to reconstitute the fuel that he has consumed."[96] In this sense men proved the more efficient machine. Mechanical motors performed work at the will of men, whereas, Gasparin argued, man's natural constitution pushed him to work.

Men must work, he believed, because their bodily hunger forces them to. "The hard law of necessity, the empire of need, and especially hunger," he wrote, "even more than the repugnance of violent punishment, oblige one to work."[97] Like Thomas Malthus, he considered the capitalist wage economy a necessary outgrowth of natural population growth. As the population increased men had to "leave the places where nature offered food" ripe for picking; men were forced to domesticate and cultivate the land. Once men had fully occupied the arable land, "new arrivals" were obliged to "obtain their subsistence by helping the landowners with their work."[98]

Gasparin regretted that inequalities were to be found between men, but considered them to "derive from the nature of things."[99] Some men were gifted with greater capacities and therefore destined to receive more than others. He compared the relative economic utility of slaves and free workers and concluded that "conceded labor" naturally presented clear economic advantages over "extorted labor."[100] The regime of wage work, for Gasparin, emerged out of the natural growth of the human population; individuals were constrained by these "natural and legitimate means" to work.[101] His social vision closely linked bodily economy and social economy.

Gasparin, like Villermé and others, feared the influence of an urban, consumerist culture on the French workforce. Urban industrial workers posed a threat to the stability of agricultural society, notably by the spread of superfluous consumer desires. When wages rose too high, decadence and indolence followed. Some workers used the excess "to deliver themselves to laziness, like Napolitan lazzarones or freed slaves." Even more pernicious, excessive wages encouraged "vanity and affluence," "that deplorable state which one sees too often among the workers of our industrial cities."[102] Gasparin feared that French workers had become accustomed to "useless and even harmful" consumer habits. "The wine drunk in the cabaret . . . gambling, clothing for men and women . . . who, on holidays, compete with each other in their level of affectation; [all these things] render wages insufficient and lead to poverty."[103] He contrasted urban workers' debauchery to the simple and righteous requirements of agricultural laborers.

In an ideal agricultural economy, wages adjusted naturally to meet basic needs. In areas "where a cold and temperate climate stimulates activity, where there is no manufacturing, where agricultural work is generalized," the price of labor would tend toward an ideal point. However, in countries like France where agriculture was not sufficiently respected, where farm work competed with poorly remunerated industrial work, "the number of workers is smaller than what agricultural production requires, and wages must be raised higher than the price of upkeep and food." Real physiological needs, the "needs of life that have limits," imposed a natural ceiling on the cost of wages. But "false needs," the desire for luxury goods, pushed wages skyward ever higher.[104] The natural equilibrium of the free market was broken.

Gasparin figured that a landowner concerned with "the acquisition of the laborer's work" must first and foremost "provide for his food, his maintenance and his replacement."[105] Replacing a laborer implied two things. First, the worker must replenish the organs and muscles that he wasted in the process of production. "As soon as a man works and becomes fatigued, one must add to the maintenance ration an amount [of carbon and nitrogen] corresponding to the loss that he must undergo."[106] Second, Gasparin contended, the employer should fund the worker's "replacement," the raising of one healthy child who could take over the job. After an extended discussion of average wages, typical budgets, and the price of flour, he set a "normal" wage for his farmworkers' families at 2.65 francs per work day, or the price of 9.82 kilograms of flour.[107]

Gasparin's model of maintenance and work rations defined needs in terms of physical expenditure. But in calculating wages he also had to account for a worker's replacement in the future; that is, for the upkeep of his children. Gasparin described human reproduction and nutrition in the same terms: "The first condition [of providing for a worker] is to provide for his food, his upkeep, and his replacement, that is to say food for his young family."[108] He struggled to define an equation for wages and reproduction. The result was a mash-up of physiological and sociological data. His wage formula was a calculus of bodily consumption, dietary standards, and social statistics. Like Dumas, Gasparin referred to the worker exclusively as an adult male. He clearly found it difficult to fit family and reproduction into his individualist, masculine, physical vision.

Gasparin's work ration was an attempt to develop a universal unit of measure for the value of labor. He calculated workers' wages by the price of food necessary to replace lost muscle and organs; that is, the quantity of matter spent in producing a given amount of labor. This measure had an economic logic. The worker's body was depicted as a "reserve" of "capital," which became depleted if the ration did not adequately replace what was spent.[109]

Gasparin thought of workers' children as "replacements," in the same way that food could replace the bodily substance that a worker lost in labor. He developed a model for wage setting based on this equivalence. The wage, in this view, should equal the cost of the replacement of the laborer's body, both his own and his children's. Wages were calculated as the purchase of bits of bodily matter.

The wage, for Gasparin, served foremost to guarantee a steady supply of workers. If wages were insufficient to maintain a worker and his family, he warned, "there would be suffering, weakening of strength, illness and finally a reduction of the number of workers." He offered a Malthusian view of insufficient wages. Agriculture would enter into a period of decline until employers again raised wages to the point where workers "could raise the children that would replace them one day."[110] In order to replace the workforce, "a family must include a man, a woman and the number of children corresponding to [infant] mortality." On the basis of average French mortality rates, Gasparin concluded that each worker's family must produce 3.22 births in order to raise one healthy male replacement. Through an unexplained mathematical operation, he asserted that "the average number of children [necessary] to maintain over twenty years will be 2.96. The wage must therefore represent the upkeep of a man, a women and 2.96 children in order to ensure, in France, the perpetuity of work."[111] This view of the wage fit neatly with the agronomists' vision of the body and nutrition. As laborers worked, they replenished their physical expenditure with food; likewise, they provided subsistence to a child who one day would replace them in the labor force.

Gasparin deduced the wages for his family of 2.96 children by use of a chemical balance equation. Women's and children's work appeared as portions of a man's work. "Numerous experiments undertaken on the consumption of men living alone or with women and children," Gasparin assured, "have proved that . . . women's diet is to that of men in the proportion 17:25." The "losses" incurred by women's sedentary work, he noted, were proportionally smaller than those occasioned by men's "violent work."[112] He provided no reference or data to support this ratio, which differs slightly from the ratio of 2:3 proposed by Lavoisier in 1791.[113] Likewise, Gasparin set children's dietary needs in relation to women's in the ratio 12:17. As children grew, their needs expanded in proportion to their productivity and wages. "The age at which young people finally earn a full salary is the age at which they attain their complete [physical] development."[114]

Ideally, Gasparin would have liked to measure the work produced by all

TABLE 3 Adrien de Gasparin, *Cours d'agriculture*, 54

	Proportions	Maintenance ration relative to weight (kg wheat)	Work ration	Total food (kg)	Value of a hectoliter of wheat at 22 fr	Value of food for the year (fr)
Man	25	0.67	0.65	1.32	0.420 fr	153.30
Woman	17	0.46	0.30	0.76	0.286	104.34
Three children	36	0.97	1.2	2.17	0.605	220.75
				4.25	1.311	478.39

members of each family, convert it to his standard unit of dug earth, and provide the rations necessary to supply that labor. "Lacking those [measures]," he allowed that one could "deduce" the family's needs from those of a single man. The base measure was obtained "from men's average weight, their maintenance ration, to which we add 45 grams of wheat for each cubic meter that a man digs in a day." Gasparin recommended that employers "multiply [the worker's needs] by 3.8, and we will have approximately the value of his labor, which represents his upkeep and the portion that he must contribute to the upkeep of his family."[115]

Gasparin attempted to render workers' nondietary expenses in equivalent dietary units. He recognized that the cost of housing and clothing varied by region and climate and eagerly anticipated social statistics on those variations. Lacking precise data, he presented a summary of his own observations, "having studied a large number of families in France." He added up a "typical" yearly cost of clothing, rent, heating and lighting, plus a small amount for "unforeseen expenses," and converted that whole sum into units of flour. Thus the yearly cost of living of Gasparin's "typical" family equaled 2,279 kilograms of wheat flour.

Food	478.39f
Rent	30.00
Clothing	100.00
Heating and lighting	10.00
Tools, utensils, unforeseen expenses	20.00
	638.39f

Representing 2279k wheat
Or, per day, 1.75f representing 6.47k wheat[116]

Here, when factoring nondietary needs, Gasparin shifted register. He turned from chemical formulae to social observation.

As Gasparin's awkward wage-setting formula shows, chemical sums did not easily convert into wages and welfare. Chemical methods produced thin definitions of need, constant over space and time. Their mobility and uniformity made them invaluable to a French state seeking to optimize its social obligations. However, these measures did not reflect historical, local, or cultural difference. Agronomists offered a limited language with which to grasp the spread of the wage relation, threats to social cohesion, and non-European civilizations. These were the objects of the nineteenth-century science of man and the social survey.

CHAPTER 4

Family, Race, Type

In 1841 the widely circulated workers' newspaper *La Ruche Populaire* warned its readers against a new brand of "scientific philanthropy." The critic, a copper turner named Schacherer, passionately condemned social investigators who presumed to measure a poor man's needs.

> Oh! Philanthropists! Your hearts are as false as the name that you give to yourselves is saintly. When will you cease to line us up like brutes, in order to keep count of our ills and our needs? When will you preach by example? What! You claim to convince us with the accuracy of your numbers, and you don't have the heart to live our lives.[1]

Despite their scientific appearance, Schacherer alleged, philanthropic surveys failed to grasp working families' existence. He suspected that social reformers manipulated their survey results to squeeze workers' livelihood. Scientific measures of need, he feared, would encourage employers to keep wages low and turn the wealthy away from charity. Schacherer's polemic reflected the rising influence of the social survey in nineteenth-century France.

We know that the social survey served as a foundational technology of the welfare state, thanks to the work of Alain Desrosières, Eileen Yeo, Katherine Lynch, Jean-Claude Perrot, and others. I argue here that the social survey emerged from a collision of biology and politics. The survey drew from the logics of biology, anthropological medicine, economic theory, scientific exploration, and worker activism. Common to all of these fields was an obsession with measuring and categorizing need. Georges Cuvier and Jean Baptiste

Lamarck defined modern zoology in a debate over the role of needs in the animal economy. The zoological conflict then profoundly influenced theories of welfare and need. Zoologists and social reformers alike asked whether needs were fixed or mutable. Need appeared as a foundational category in both natural and human sciences.

Social surveys supported two opposing ideologies of social reform and of wage setting. One, characteristic of Joseph de Gérando and Frédéric Le Play, employed surveys to divide the French population into social-medical types. Drawing from Georges Cuvier's zoological theory, they classified their subjects by fixed and immutable categories. These surveys replicated the transactional format of chemical studies. They examined individual subjects, their inputs and outputs, at great depth. Their recommendations were put into practice by private charity organizations and para-political associations like Le Play's Society for Social Economy.[2] Their surveys created the categories that would undergird early forms of French welfare, particularly family allowances.

Another, progressive use of surveys was developed by socialist economists and workers' associations. Workers' surveys rejected fixed types in favor of a language of transformation and progress. They propounded a statistical approach that would capture different people's needs as they expanded under changing economic and cultural conditions. Needs, for progressives, were not predetermined but rather constantly negotiated. Workers' efforts to aggregate wage and budget data suggest that already in the 1840s, some workers sought to establish class identity and power through statistics. Eileen Yeo has observed that in England, Chartists collected statistics on wages and the cost of living as early as 1839.[3] Statistics formed part of a working-class culture well before the popular spread of positivism and well before the International Working Men's Association endorsed statistics as a tool for proletarian liberation. Workers wielded data on wages and the cost of living in support of demands for a legal minimum wage.

The social survey responded to a need to make sense of a world stripped of feudal custom. At the close of the eighteenth century the French monarchy lost its prerogative over social knowledge. State censuses of national population and economy, which the court had censored from public knowledge, were supplanted by private efforts. Local academies, engineers, agricultural and medical societies gathered and published figures on population and commerce. The Revolution completed this transformation. Revolutionary governments set about dividing the territory into equivalent units and classifying the national population by property, land tenure, and revenue.[4] Civil society emerged as an entity distinct from the state, understood as an aggregate of individuals, each with their own property relations, interests, and relationships.[5] With the abolition of royal privileges in 1789, social distinctions became a question for scientific investigation.[6]

The age of scientific exploration, which peaked in the eighteenth century, presented Europeans with the conundrum of cultural difference. European encounters with foreign peoples raised the problem of relativism and universality, how to define a common humanity while accounting for distinction. Explorers and their armchair counterparts established scientific systems, typologies, and hierarchies for analyzing culture. Even as they sought to comprehend distant peoples, scientific explorers turned their eyes inward toward France. Anthropological notions of culture, civilization, and "natural man" applied as much to rural or poor French citizens as to aboriginal peoples. Manuals for overseas exploration were written with an eye to possible application by French social investigators.[7] Theories of environmental determinism, heredity, and the temporal march of progress reflected back upon France itself. In this context "society" and "culture" ceased to appear as given and became objects of scientific investigation. The social survey was a form of political theory as much as a record of observation.[8]

Social surveyors practiced thick description. They burdened their reports with numbers of all kinds, corresponding to multiple forms of social activity. They recorded observations of lifestyle, social ties, and material conditions; they tied their subjects closely to physical conditions and social milieu. They applied methods of budgetary accounting, natural scientific description, and clinical examination. The most widespread form of survey, and the one under consideration here, was the family budget. Budgets recorded a single family's consumption and production, including all forms of income and expenditures in a unified framework.

Surveys anchored their subjects in local and particular cultural difference. A range of "anthropological" sciences coalesced around a common program of social observation and regeneration.[9] Surveys associated physical health, economic status and cultural advancement, as well as moral, political, and physical conditions. Surveyors classified their subjects by family status, property and revenue, and nation or locality. Race, sex, and class were bound together under this general framework.

WELFARE AND COMPARATIVE ZOOLOGY

The object of Schacherer's ire against "philanthropic" social investigators was Joseph de Gérando's 1839 work *On Indigence*. Schacherer chose his target well; Gérando was perhaps the most influential originator of the modern French social survey, and one of its more conservative practitioners. His work crossed philosophy, scientific exploration, social investigation, state administration, and private charitable philanthropy. In the history of anthropology Gérando is known as an early proponent of field observation;[10] in the history of welfare he appears as an inventor of the home visit.[11] I argue here that the two are closely linked.[12]

Gérando modeled his "human science" on a particular form of natural science, that is, his friend Georges Cuvier's comparative zoology. Gérando gathered numbers and descriptions of poor and exotic peoples in order to define fixed sociocultural types. This methodology became the basis for a characteristically French style of social survey. The influence of comparative anatomy on French social surveys lasted into the late nineteenth century. Gérando's formulation heavily influenced Frédéric Le Play, who is often identified as the originator of this method. Le Play likened his work to that of the zoologist—clearly referring to Cuvier—who compares the anatomy of different species.[13]

In the background of the modern French human sciences was an argument about humans' potential for progress. How fixed were the divisions between different classes of people? Could "savages" be brought to the level of French civilization? How different were humans from other animals? Should the human sciences seek to classify fixed types or to record constantly changing relationships? Were human needs relative or universal, and how could one know? This argument was expressed perhaps most powerfully in the debate between Cuvier and his colleague Jean Baptiste Lamarck. Followers of the Cuvier-Lamarck quarrel understood that although it dealt with animal anatomy, it held deeply social implications.

Cuvier built his scientific reputation on meticulous comparison of anatomical characteristics and on a vast catalog of observation. Animal physiology, he claimed, could be grasped only through exhaustive comparison of organs from many different species. Only by studying multiple examples of a specific organ could one begin to infer its function in the animal economy. An animal's bodily organization was entirely determined by the relationship between its organs' functions, which Cuvier called the "conditions for existence."[14] In order for an animal to exist, its bodily functions and organs must be compatible.

Cuvier classified animals according to their functional relationships. Bodily organization always corresponded to a given set of functions. If an animal's body were dominated by one kind of organ, the naturalist could predict the size and importance of other, correlated organs.[15] Meat eaters with strong jaws must always also have sharp eyesight and smell, rapid movement and clawed

FIGURE 4. Paris, France: 1800. Anatomy lessons of the baron Georges Cuvier (1769–1832), French naturalist, in the Jardin des Plantes of Paris (Museum of Natural History). French lithography, 19th century. © Jacques Boyer / Roger-Viollet / The Image Works.

feet. Pointed teeth could never be found alongside a horned foot.[16] Thus sprang the legend that Cuvier could reconstruct an animal's entire body on the basis of a single bone. Animal types corresponded to functional groupings. Mammals, reptiles, and birds could be differentiated by the speed of their breathing and the quickness of their movements.[17] An animal's scope of activity, therefore, must be inferred from observing its physical organs. Bodies were optimally arranged to respond to differing conditions of existence.

For Cuvier, the animal economy was immutable. An animal's needs, he wrote, were fixed by the law of nature. "It is one of the conditions of existence for each animal, that its needs are [always] proportional to the faculties that [the animal] has to satisfy them."[18] The scope of an animal's existence, the range of its needs and actions, was predetermined by its physiological organization. Just as the naturalist could deduce that flat molars always accompany horned feet, he could predict the extent of an animal's needs from its organs. Needs were set by the laws of physiology and the conditions of existence.

If the body's physical nature determined a person's needs, then physical differences could imply a greater or lesser level of need. Doctor Pierre Cabanis offered a physiological justification, based on the weakness of female muscle fibers, for the inequality of the sexes. This handicap lent women to sedentary work in the protected interior of a home and thus to a different and sexually specific "physical and moral system."[19] In an analogous manner, climate served to explain different nations' needs. Cabanis and other practitioners of "anthropological medicine" sought to classify the population according to a complex grid of difference. Individuals could be grouped, using a combination of clinical and statistical methods, into "types."[20] Temperament, temperature, age, sex, and disease all worked to alter the body's organization and thereby its mental, moral, and social capacities.

Cabanis's contemporary Xavier Bichat explicitly suggested that each individual possessed a fixed and limited potential. People had a certain vital energy or sensibility at birth; all one could do was to maximize one's performance within those limits. Bichat deduced a moral and social order from this physical law, whereby some were destined to achieve greater works and higher satisfactions than others.[21] Bichat's theory reflected a common consensus among early nineteenth-century medical reformers that individuals held unequal needs and capabilities.[22] Medical authorities published numerous manuals and popular guides designed to propagate bodily "education" and discipline. Strict regimes were imposed in clinics and hospitals. Sexual norms and women's activities were subject to particular scrutiny. Proponents of hygienic "regeneration" sought to reinforce social hierarchies that had been challenged by the Revolution.[23]

Where Cuvier saw fixed and stable types, his contemporary Jean Baptiste Lamarck saw mutability and progression. In Lamarck's view, an animal's needs stimulated its vital energy and shaped its physical body. Environmental changes sparked new needs, which influenced the shape and size of an animal's organs. New needs were the prime cause of growth and change in the animal economy.

Cuvier grounded his analysis in bodily functions; Lamarck's first principle of anatomy was an animal's needs. "Circumstances are causes that bring about new needs, needs cause new actions, repeated actions create habits and tendencies." Scientists must understand, as "objects of the greatest importance," the dynamic interactions between an animal's environment and its needs. Needs shape the scope of an animal's movement and its basic composition. As a result of new habits, "one or the other organ . . . [is] more or less used," vital fluids concentrate in new areas of the body, and organs gradually change over time or over generations. This is a process of transformation and improvement. Nature is constantly "conserving and perfecting everything that has been acquired in [animal] organization."[24]

Lamarck explained the flourishing of the human race by the same process of expanding needs and improving capacities. Vertebrate animals responded to four primary needs: to eat, to reproduce, to avoid pain, and to seek pleasure and well-being. All animals' habits and instincts derived from those needs.[25] Early humans, Lamarck argued, mastered those basic needs and began to develop new, more complex needs. Having dominated other animal races, humans seized more space to expand and "new needs were successively created that excited [man's] industry and gradually perfected his faculties and capabilities."[26]

Lamarck, like many of his contemporaries, identified need as the moving force in human development. Humans "progressively acquired modifications in their organization" in response to new needs, most importantly the ability to use language. The need to communicate ever more complex ideas led humans to exercise their vocal abilities, develop their tongue, throat, and lips, and articulate sophisticated sounds and words. "In this respect, needs did everything: they gave birth to [new] efforts, and organs . . . developed through habitual use."[27]

Cuvier used his institutional prestige to denigrate and mischaracterize his rival, whose historical reputation suffered badly. However, early nineteenth-century social reformers attended to Lamarck assiduously. Numerous doctors attended Lamarck's course at the Museum of Natural History, including Louis René Villermé and future socialist politician Philippe Buchez.[28] Lamarck's influence showed in the work of progressive social economists. His anatomical lectures reinforced the notion that environmental circumstances determined a person's physical ability and well-being. He imparted to his auditors faith in the human potential for infinite advancement, both cultural and physical.

In the debate between Cuvier and Lamarck, Gérando fell solidly on the side of the former. Gérando, like Cuvier, came of age in the post-Revolutionary upheaval and catapulted while still young to the summits of the new intellectual elite. Gérando's young career demonstrated a knack for opportunism and lucky patronage. Originally destined for the clergy, he fought against the Revolution in Lyons; then, having escaped capture there, he fought his way back into France as a member of the Revolutionary army.[29] He entered Parisian society under the wing of Germaine de Staël and Lucien Bonaparte, who helped both Gérando and Cuvier obtain entry into scientific societies and government sinecures.[30] Gérando, who refused to take a position either for or against Condillac's sensationalism, became a consensus candidate for prize honors and then for election to the new Institut National.[31] In that same year, 1804, he was hired in the Interior Ministry, under the direction of Lucien Bonaparte, and gradually worked his way to become a counselor of state. Although he first made his name as a philosopher, his role in the Napoleonic elite was to synthesize and reform. He was active in reforming medical education and in founding the Royal Academy of Medicine.[32] After Napoleon's fall from power, Gérando vigorously pursued private charitable operations from children's asylums to convalescent workhouses to philanthropic societies.[33]

Soon after arriving in Paris as a young man, Gérando joined Cuvier as founding member of the new Society for the Observation of Man. The society's mission was as wide-ranging as the disciplinary specialists who composed it: medicine, deaf-mute studies, special education, anatomy, and philosophy. Established by two conservative Catholics, the society also participated in the counter-revolutionary movement of Directory and Empire intellectuals. Its founders envisioned a science of man that could stabilize and regenerate the French population. Like many conservative intellectuals in the wake of Revolutionary upheaval, they sought to reinforce individual self-discipline, paternal authority, and traditional social divisions.[34] The society is best known for scientific instructions, which Gérando and Cuvier wrote in 1800, destined for Nicolas Baudin's expedition to Australia and the Pacific Islands. Cuvier's instruction focused on the physical observation of native peoples, while Gérando's covered the linguistic, social, cultural, and economic fields. George Stocking identified these two instructions as landmark documents in the history of modern anthropology.[35]

Gérando's instruction to the Baudin expedition explicitly couched his social-scientific method in the terms of Cuvier's comparative anatomy. "The science of man is also a natural science, a science of observation," he wrote, by which he meant a science that "assembles facts in order to compare them. . . . Natural sciences are only a sequence of comparisons."[36] Gérando conceived of his human subjects as exemplary types, which could be mapped on a stable grid of analysis in the manner of invertebrates or fish fins. Native peoples offered an ideal territory, in his view, to discover and compare human types. "Of all the terms of comparison that we might choose [in the human sciences], none are more curious, none stimulate more useful deliberation than savage peoples."[37] Only savage peoples living in a "natural state" could reveal the fundamental causes of human nature, unadulterated by history.[38]

Comparison, Gérando believed, would capture the essence of what distinguished one person from another. Like Cuvier, he sought to separate

FIGURE 5. Newspaper of Nicolas Baudin (1754–1803), French sailor and explorer, who received from the French Directory the command of two corvettes (*Geographe* and *Naturaliste*), with the mission to investigate the coasts of the New-Holland (Australia). © Roger-Viollet / The Image Works.

"secondary circumstances" from "primary and fundamental circumstances." Some qualities were basic to human nature; others were residues of culture or happenstance. First causes, alone, "belong to the principles of existence."[39] Gérando's language again echoed Cuvier, who classified animals according to his principle of "conditions of existence." ("Since nothing can exist if it does not meet the conditions which make possible its existence, the different parts of each being should be coordinated such that the total being becomes possible.")[40] The organization of each being corresponded necessarily, and unalterably, with the nature of its environment. For Gérando, primary causes were "climate, [bodily] organization, and the habits of physical life."

Gérando imagined that his method would lead to a tableau of "natural varieties"—again, a zoological term—of humankind. This was the basic operation upon which relied all of his future scientific and philanthropic work. Having defined the human sciences' "essential laws," he could turn them into an empirical tool of measurement. "Here," he proclaimed, "we can find all of the material we need to compose an *exact scale of different degrees of civilization*."[41]

Gérando ranked his human subjects on a universal scale derived from natural observation. Each level on the scale represented a distinctive set of "needs, ideas and habits."[42] Gérando's scale mapped all possible human "conditions of existence." Its gradations served to define what he called, at various times, "primitive units" or "normal types." A person's place on the scale determined his or her potential to need, to act, and to think. Gérando carried that scale of civilization with him for four decades.

The scale of civilization corresponded to a systematic set of criteria. First, the observer of peoples should learn and analyze their language. Next, he should select an individual subject and record the "circumstances of his physical existence," which corresponded to primary needs. Like Cabanis, Gérando considered sensations of need as fundamental to human existence. "What is the intensity of the Savage's hunger, thirst and fatigue; what are the effects that determine these needs?"[43] Primary needs depended on both the surrounding environment—the available food and water—and an individual's physical capacity, his "penchant for laziness" or movement. These physical conditions represented a people's "primary and fundamental circumstances," its "principles of existence." Gérando's "primary circumstances" echoed contemporary language used to describe paupers: by their needs and industriousness.

A people's level of civilization depended on mastery of its physical conditions of existence, which allowed secondary, "reflective needs" to develop. Secondary needs included things like curiosity, surprise, fear or uncertainty, and attraction to amusements or pleasure. Strikingly, they also included the subject's awareness of his own subaltern status, "the idea he might have of his inferiority and of the development that he might have had."[44] Having mapped out the range of his subject's needs and placed him on the scale of civilization, the observer should then attend to other factors that might differentiate between individuals within a single society. Many samples were necessary to establish the full range of "varieties": age, sex, temperament, and "the organization of circumstances" (by which perhaps he referred to social status). Thus the observer, by detailed comparative work, grades his subjects by the scale of civilization.

Gérando was not interested in knowledge of exotic peoples for its own sake. Rather, he used his instruction to develop a disciplinary system of classification. He offered a holistic vision of human science in the service of a utilitarian end: his method was designed to promote the interests of the French state. Scientific knowledge served as a means to extend the reach of French commerce.[45] Even while composing his instruction, he had his eye on the census being carried out by the post-Revolutionary French state.[46]

With his instruction Gérando bridged Enlightenment universalism and the nineteenth century's more pragmatic utilitarianism. Systematic surveys of the French territory, regional catalogs of natural, economic, and cultural resources, had begun to appear in the late eighteenth century. Revolutionary state authorities turned this library of descriptive "medical topographies," "dictionaries," and "tables" into instruments of governance, via an expanding network of local prefects. Surveyors carried out "an encyclopedic project, seeking to build a general science of nations [*peuples*] through the unity of [natural, social, political, and cultural] knowledge."[47] Some of these included sample family budgets, as a measure of a people's well-being. In the early nineteenth century surveyors debated the utility of the descriptive form over quantitative mathematical tables. Proponents of descriptive statistics referred to Gérando's Baudin instruction as a model for state surveyors.[48]

In 1800 Lucien Bonaparte created a statistical office within the Interior Ministry, charged with collecting local prefects' responses to an ambitious national survey that had begun under the Directory. Gérando's instruction represented a bid for a position within the network of scientific surveyors and, more prosaically, a job. Through the auspices of the Society for the Observation of Man, Gérando ensured that his instruction received widespread publicity. This strategy bore fruit: Bonaparte hired him.

After Napoleon's fall from power, Gérando lost political favor and turned to the private sphere. During the Restoration and July Monarchy he honed his method for use by philanthropists. Twenty years after the Baudin expedition, Gérando wrote the work for which he is now best known, *Visitor to the Poor* (1826). There, and in a subsequent work entitled *On Indigence* (1839), he proposed a method to measure indigent people's needs and manage poor aid. He addressed these manuals to philanthropists seeking a rational scientific method. Just as the earlier instruction had done for the scientific explorer, *Visitor to the Poor* guided the charitable investigator. Here Gérando found his calling and a forum, one which would earn him historical notoriety. He took the "scale of civilization" that he had developed in his anthropological instruction and applied it to poor Frenchmen.

In the philanthropic manuals Gérando's scale of civilization became one of "degrees of misery."[49] Here again these were measurable gradations, based on empirical first principles. In entering the home of an Australian aborigine or a poor French family, the observer "penetrated" to a primal stage of human development.[50] There it was possible to establish a measure of human needs and capabilities. Gérando directed the philanthropic observer, like the explorer, to begin with a study of a healthy adult male. Given that baseline, he could then derive different varieties according to age, sex, health, and temperament. As he had in his instruction, Gérando distinguished between "primary necessities," including food, clothing, furniture and housing, and "social conventions" and "habits" reflecting the level of cultural development.[51] Both must be accounted for in order to set a "just measure of real needs."[52]

Gérando's scale served two uses: "the classification of indigents, and the evaluation of the degrees of misery."[53] He ranked charity cases on a scale of development and indigence. More pragmatically, each level on the scale corresponded to a specific set of needs—and thus to an appropriate measure of welfare payments. With rare exceptions (such as wealthy widows fallen on hard times), his measure could be applied uniformly to all cases; it was a "general formula" for evaluating need.[54] Gérando recommended his scale for use in charity and in setting workers' wages. Philanthropists, with the scale in hand, could "make a judgment about real needs" and distribute payments on a "positive basis."

Visitor to the Poor included a standard survey form with which to measure each subject's needs and means. One side listed information about civil status, health, family size and age, employment, revenues, and willingness to work. The form's back side enumerated consumer goods—housing, furniture, clothing, and food—at the subject's disposition. As his anthropological instruction directed the order and priority of scientific observation, so did his philanthropic manual: "If we take absolute pauperism as a unit, you will have three different *primitive units*: one for a man, one for a woman, and one for a child. This unit will represent everything that the unfortunate [subject] needs each day, all included."[55]

TABLE 4 Joseph Marie de Gérando, *Le visiteur du pauvre*, 140–42

Endiameter

1st Part

1. Name (first and last) of poor
2. Residence (current/previous)
3. Sex
4. Age
5. Profession
6. Single/ Married/ Widowed
7. Children (Number, Sex, Age, Civil status)
8. Parents
    a. (Father and Mother, Brother and Sister, Close relatives)
    b. (Residence, Condition, Help that he could receive from them)
9. If he is infirm (Which infirmity, Curable or Not curable)
10. If he is sick (Which sickness, Since when, Which doctor is treating him, Does the dispensary admit him)
11. If he is able to work (Does he have work, Is the work sufficient, For how long)

2nd Part

The visitor will indicate what the poor man has and what he is missing

1. Rent, annual price
2. Bedding (Mattress, Sheets, Blankets)
3. Furniture
4. Clothing
5. Underwear
6. Shoes
7. Heat
8. Food
9. Medicine
10. Children's layette
11. If the children go to school (What school, Free of charge or Paid, Price per month)
12. Type of work
13. Profession or other means of working
14. Debts
15. Belongings at the Mont de Piété
16. Apprenticeship of children (With whom, For which type of work, Under what conditions)

In his later years Gérando reworked his notion of a "primitive unit" (with its evident reference to "savage peoples") in light of statistician Adolphe Quêtelet's "normal man." Gérando adopted Quêtelet's terminology; the "primitive unit" became a "normal type." Gérando suggested that charity administrators establish a "normal type" for male workers, then for others according to sex, age, temperament, profession, location, and state of health. He labored to bridge the Enlightenment descriptive method and modern statistics. He saw his "normal type" as a mean, around which individual cases would converge. In this sense he was not far from Quêtelet. Both saw their "normal type" and "normal man" as entities representing a whole group. They sought to establish convergences, not individual variations.[56]

A tension appears between Gérando's desire to standardize and his attachment to the empirical method. It is never quite clear whether his scale of misery exists prior to observation or is a final result of accumulated knowledge. He claimed scientific status on the basis of Cuvier's comparative method. Given data on primitive peoples, Gérando believed that he could derive a scientific measure of "primary and fundamental circumstances," or basic needs. This measure would be refined with further observation. He proposed that local charity offices be established to collect philanthropic investigators' standardized forms, from which they could draw general, "more positive" conclusions.[57] Yet in its application the measure was disciplinary and inflexible. Gérando, following Cuvier, believed that repeated observations would converge around a set of fixed types. Each type was determined by a specific set of circumstances, "conditions of existence," which remained constant. Thus poor aid and wage rates, too, should converge around an unchanging scale of values.

In his later work Gérando employed both statistical data and case studies, including wage rates and typical family budgets. "One can see how difficult it is to evaluate [needs], how impossible it is to assign them a universal and constant value," he admitted. "[However] a more certain calculation can be established, on which we have chosen to base our definition of indigence."[58] *On Indigence* contains twenty pages of statistical data. Gérando recorded typical daily budgets for working families in various regions and countries, a prisoner, a soldier, and a hospital patient. He published a table of average wages in each of France's ninety-one departments and composed a catalog of "indispensable expenses for a family composed of father, mother and three children." One typical budget was for "workers in large cities," and another, diminished, gave a set of figures for country folk. According to his calculations, city dwellers needed, at minimum:

Food.	
Bread at 16 centimes per day [per person].	
Per year	296 fr 40 c
Meat, eggs, cheese, vegetables, seasoning at 50 c per day.	182,50>
Fermented drinks, at 25 c per day.	91,25>
Housing.	
Dwelling	50 fr>
Fire and lighting	40
Contributions	10>
Replacement and maintenance of furniture	30>
Clothing.	
Father	50 fr
Mother	30
Three children	60
TOTAL	840 francs 15 centimes[59]

In his treatment of statistics, too, Gérando bridged an Enlightenment spirit of inquiry with nineteenth-century discipline. Having accumulated a mound of disparate statistics, he decided that his "normal type" in fact should be based on current wage rates: "The wage rate earned by the least well-off independent worker will serve as a point of departure to express the total needs of a healthy adult. Any deficit [from that amount] will measure the degree of indigence. This calculation will vary according to age, locality, health, profession; for each of these classes we will take a normal type as our regulator."[60] Gérando's natural scientific, empirical edifice ultimately served to hold up the prevailing wage structure.

FAMILY AND RACE

Gérando introduced Cuvier's comparative method to the human sciences. Human types, like animal species, were fixed by the conditions of their existence. Gérando instructed his observers to penetrate deeply into the physical and cultural existence of a few select individuals. By comparing results through a standardized grid of analysis, he situated his subjects on a hierarchical scale of civilization. Gérando's anthropology described "peoples" but not races. His concept of human "varieties" was individual and corresponded to each subject's age, sex, race, and temperament. When he dealt with families in his philanthropic work, Gérando treated them as a numerical sum of individuals. His "family budgets" were simple additions of each individual's revenues and expenses; he engaged with family groups only to note that collective purchasing could lead to some cost savings. In the second half of the nineteenth century, the status of the family changed. Cuvier's fixed types and Gérando's philanthropic method became the foundation for a racial theory of need. The family budget, in the hands of Frédéric Le Play and Emile Cheysson, came to represent a racial scale of civilization.

Le Play and Cheysson employed a systematic, quantitative method drawn from the world of engineering. Like Gérando, they practiced a form of ethnographic observation in which the investigator "dissects [his subject] to the bone and penetrates the secret of his situation."[61] However, they refused to consider individual subjects in isolation. To them an individual was entirely determined by his or her conditions of existence and, above all, inheritance. The family was their fundamental unit of observation, their "social molecule."[62] Families, for Le Play and Cheysson, were windows into the fundamental characteristics of a race. Family and race set the horizon of human needs and capabilities.

Le Play began his career at the elite Ecole des Mines under the influence of Saint Simonian thought. At the close of his studies in 1829 he toured the mines of central Europe with Saint Simonian Jean Reynaud; in the coming decade Le Play and Reynaud worked out their social and scientific ideas in tandem. They shared an interest in miners' craft knowledge, working and living conditions. Le Play concluded from this early tour that some mines managed to withstand price-slashing competition only because of their managers' concern with workers' well-being. Where mining companies supplemented their workers' pay with charitable donations, workers maintained their commitment and productivity.[63] On the basis of such observations, while traveling to mining sites across Europe, Le Play developed his method of "family monographs." By the 1850s, social investigation had displaced mining as his primary occupation. His *Ouvriers des deux mondes* (1857–62) contained four volumes of case studies on workers across Europe, each with reams of close description and budgetary figures. Each chapter described a family's intimate life, its possessions, its religious practices, social networks, daily habits, and means of survival in exhaustive detail. He applied rigorous double-book accounting methods and a deep concern for the bonds that anchor people in place and time.

As he aged, Le Play became increasingly nostalgic for a lost age of agrarian economies, customary practices, and paternalistic authority. Since the French Revolution, he believed, modern Europeans had lost their roots. Urbanization and industry had rent asunder the timeless traditions upon which prosperity and peace depended. Social ties had loosened and custom lost its sway. Constant strife between social classes resulted when "customary" proportion between needs and pay rates was broken.[64] In this context he turned toward Cuvier and Gérando's notion of conditions of existence. Their comparative method permitted the observer to fix his subjects on a stable grid of analysis. Revolution had unmoored Europeans from their customary networks; Le Play sought to bind them within a thick fabric of detail. Like Gérando, Le Play looked to primitive peoples to reveal a true human nature, unadulterated by culture or history. "Fixed types," for him, was a political and scientific principle.

Le Play explicitly drew upon the comparative anatomical method. He defined "the laws of social science" on the basis of "the simplest cases," apt to reveal primary and fundamental causes. Thus peripheral, rural populations served the same purpose for Le Play that savage peoples had for Gérando. Their lives reflected the true principles of human nature. Social science was not served well by abstract theories or mathematical tables. Instead, scientists must undertake systematic and minute observations.[65] Le Play compared his method to that of "a zoologist who, to describe a living species, applies anatomical and physiological procedures of investigation to a few [selected] individuals of that species."[66] Social science was necessary for the art of government, he argued, in the same way that global natural histories depended on the detailed observations of anatomy and physiology.[67]

Le Play's specimen, his basic unit of study, was the family. Where Gérando described individual people as quasi-biological "varieties," Le Play applied the same language to families. When the social observer observed like a zoologist, his lens captured "the family, the true social unit."[68] Facts about families formed "a basis for statistics, public administration and industrial organization." Even more important, Le Play argued, family life contained "the true principles of social science." Only by studying families could the scientist understand the fundamental bases of social order: the influence of religion, the fecundity of marriages, inheritance regimes, property arrangements, and the relationship between individuals and collective bodies.[69] Families were Le Play's "species individuals."

Like many conservatives, Le Play turned to the family as a source of stability and determinacy in an age of upheaval. The family, for social Catholics and hygienists, held the key to social regeneration.[70] The family household were the subject of civil law, the recipient of wages, and the payer of taxes. Early nineteenth-century tax censuses, which measured household income, provided social economists with much of their raw data about family budgets. Households, and houses, were also the focus of intense hygienic attention and reform. Both Le Play and Cheysson actively promoted (and in Cheysson's case, developed) property ownership for working families, as a moral and material anchor.[71] Fears about the family amplified as social and economic conditions appeared to loosen its integrity. Common law households were the norm among urban workers in the nineteenth century, raising fears about promiscuity and illegitimacy.[72] As individual wage contracts began to replace family working arrangements, reformers worried that wages would not guarantee the reproduction of labor.[73] Families appeared as a "natural group" that could shelter its members from the deleterious effects of urban, industrial life.

Le Play, like Gérando, placed the highest importance on basic needs. Le Play's observations focused on a people's means of subsistence, its "material and moral conditions of existence."[74] His family budgets recorded every effort, every transaction, every acquisition in a family's daily life. He meticulously divided this budgetary data into minute typologies. "Resources" included wages, garden products, communal property; "expenditures" ranged from food, housing, and clothing to insurance, religious expenses, education, and "improvident" spending on cabarets or makeup, to name only a few examples.[75]

Le Play claimed to attend to "the fundamental details of [his subjects'] existence, in the same way that biologists report consistency and regularity among the individuals of the same species."[76] These regularities revealed what Gérando called the "circumstances of physical existence," the fundamental causes of human nature. Like Gérando, Le Play believed that a people's environment fixed the range of its potential development. He classified human types according to their means and their needs.

> The combinations that different races of men use in order to provide for their existence vary according to the nature of the soil, water and climate. They offer, to the observer of these races, the most apparent and distinctive traits. The nature of efforts that individuals undertake, to ensure their subsistence, in many ways determines the rank that their race occupies in the hierarchy of societies.[77]

Le Play fashioned Gérando's scale of civilization into a racial hierarchy. In the last years of his life, he extended his unit of analysis from family to race. A "simple race," he suggested, was nothing more than "a large family composed of households issued from the same ancestor and sharing the same name."

Historically, as families from several ancestors shared the same geographical area, "complicated races" of "entanglements of small families" emerged. Over time, family bonds were rent by geographical distance and the division of labor. The fading power of the patriarch was supplemented by new institutions, religion and the sovereign state.[78] To study a simple race, Le Play suggested, one need only observe a single family. Social classes did not appear among the simple races. However, in modern Europe "complication is the characteristic trait of all societies" and the scientist must "multiply his observations" to capture representative families from various classes in each society.[79] Thus a collection of family budgets revealed the nature of a race.

Le Play offered a historical account of the development of race, grounded in geography and ancestry. His student Emile Cheysson integrated this racial hierarchy of civilization into late nineteenth-century racial theory and eugenics. Cheysson followed Le Play's path through the Ecole des Mines, where he mastered modern statistical methods. He applied these techniques to social questions via a kind of "econo-engineering."[80] Cheysson was a central figure in the fashioning of social policy, as one of the architects of modern French workers' housing and family allowance programs.[81] He also undertook extensive family monographs. Like Gérando and Le Play, Cheysson sought to describe fixed types, which he defined as specimens reflecting the average of a given population. "The type," he wrote, "is what constitutes the true essence of the monograph."[82] He likened his monographs to an "ethnographic collection," juxtaposing "types of various races in order to compare them trait by trait."[83]

Cheysson stressed the role of inheritance in Le Play's schema. Le Play considered regimes of property inheritance as one of the seven pillars of social science. Cheysson combined Le Play's economic notions of inheritance with the "the law of heredity." "Each one of us," he wrote, "comes from both our parents and our race." In an eccentric reading of Francis Galton, Cheysson concluded that the relative part of "race" in genetic inheritance could be quantified. "Physiology" proved to Cheysson that "our nation [*patrie*] is our second mother."[84] This was a late nineteenth-century version of the principle of fixed types. Cheysson bridged Cuvier's theory of fixed needs, Le Play's social Catholic conservatism, and modern eugenic thought. Human needs and capabilities were set by racial ancestry.

SOCIALISM AND STATISTICS

Gérando and Le Play employed the family budget as an instrument of a conservative human science. They used Cuvier's conditions of existence and the medical model of fixed types to build a universal scale of civilization. Their contemporaries were quite aware of the political implications of a fixed scale of needs. If a worker's potential was fixed by custom and inheritance, little grounds remained on which to claim a better standard of living. Workers like Schacherer recognized this danger and issued public refutations.

Against Cuvier's fixed types they offered a progressive, historical vision of human needs and abilities. In this, they drew from the biological principles of Cuvier's rival Lamarck and the radical economics of Jean Charles Léonard de Sismondi. They argued that conditions of existence evolved as a result of changing economic conditions. Needs depended on the level of culture and well-being of a given age. This position carried moral and scientific consequences. First, it was incumbent upon society to raise that level. Second, only a moving statistical measure could capture the constant evolution of workers' needs. In place of fixed types, they called for regular collection of local statistics.

Schacherer, in *La Ruche Populaire*, warned that Gérando's "scientific philanthropy" was neither scientific nor philanthropic. "So-called philanthropists" designed scientific family budgets to squeeze laborers for more work and less money. Schacherer foresaw that Gérando's measures would serve first and foremost "to harden the hearts of the prosperous." If those figures became a commonly accepted standard, they would justify depriving impoverished workers of all but the most basic goods. "Such unfounded [figures] lead a certain municipal council to say that there are no poor people in France, and that the word 'pauperism' was invented by [political] parties to horrify those in power."[85] With Gérando's measures in hand, municipal aid offices would claim that no one in their district qualified as a pauper. Entrepreneurs would invest more money in the stock market "instead of helping the workers."[86]

Despite his unqualified condemnation of Gérando's budgetary exercise, Schacherer nevertheless felt obliged to set the statistical record straight. "May I ask the Baron what foodstuffs, meat or vegetable, allow five people to live by spending only ten centimes per person each day in the city, and only five in the countryside? . . . Truly, Monsieur le Baron has been dreaming and we workers would be very pleased if he would be kind enough to give us the means to live so cheaply."[87] The author himself turned to typical budgets in order to prove that workers deserved a better everyday life.

By the time he had finished, Schacherer had compiled one of the most extensive lists to date of expenses and wages in various towns and countries. He offered his own version of a worker's budget taken from "the least-paid class." He revised quantities and prices for all of Gérando's categories and included one more: "Obligatory rest days and holidays (which no philanthropist speaks about) . . . 215 fr 60 c." Schacherer's lowest-class worker's family lived for 1,367 francs per year, a considerable jump from Gérando's calculations.[88] Schacherer figured that a working family might earn just enough to meet this standard, but only if no little extras, accidents, or dependencies intervened to ruin it. Even given his higher budget, "this rabble is not supposed to drink wine or enjoy itself; it can't be unemployed . . . or sick."[89] He derived his measures from a sample budget of a family in Nîmes, first published by the Saint-Simonian workers' journal *L'Atelier*. *L'Atelier* in the 1840s campaigned in favor of a family breadwinner wage. Its editorialists referred to family expenses as a rhetorical device to argue for higher wages in the service of a socially useful calling: the raising of children to replace their parents in the workforce.[90] Despite his scathing critique of Gérando, Schacherer himself adopted family budgets and a quantitative method to refute his adversary's claims.

The article produced prodigious tables of workers' diets and wages in ten European countries, Russia, and several American states. "Everywhere [workers] are bent under iron rods . . . everywhere they are unknown and despised; but let us hope, workers, let us hope, to soon end the ills that overwhelm us, to have at last our share of well-being." A common desire for well-being, Schacherer speculated, would unite workers of all nations and trades to fight for higher wages. "Let us gather our forces, associate our ills, for a pain that is shared has a weaker hold on our souls."[91]

For Schacherer and his colleagues at *L'Atelier*, family budgets served to document workers' share in the social distribution of wealth. Scientific measures of consumption, like the family budget, did not capture any inherent qualities or types. No single budget told the whole story; conditions changed over time and space, and must be captured through statistical measures. As progressive social economist Eugène Buret put it in 1841, "Populations themselves create their own needs. . . . The economist can only make note of a people's needs, as they have been created . . . on the basis of an average of the needs of the people he studies."[92] Workers' access to wealth and health depended on a shifting balance of power. This view of needs and well-being echoed Sismondi's economic theory.

In a class-based society like modern France, wrote Sismondi, "the needs of a worker are necessarily strongly constrained." Needs were neither fixed nor equal. Modern industry and productivity, he submitted, could potentially summon "the forces of the whole society" to provide for the well-being of all citizens. But in a profit-based economy, "efforts today are separated from their reward": one man works and the other relaxes.[93] The social division of labor did not derive from any natural principles, but rather from a specific political economy. An individual's needs depended on a set of concrete conditions, notably his or her place in the social order.

Workers were constrained to enjoy only basic needs; property owners benefited from a wider range of complex needs.[94] Sismondi himself recoiled from the notion that all Frenchmen should be put to work; like the classical economists he believed that luxury and profit were necessary economic stimulants. He pushed employers to accept responsibility for their workers' medical care, childcare, and old age pensions. His *New Principles of Political Economy* (1819) stimulated others to more radical claims.

Sismondi contributed to the denaturalizing of economics. He derived economic principles not from physiology or natural history but from the history of man. For Jean Baptiste Say and David Ricardo, economic things were "contiguous with physical nature."[95] Sismondi broke that continuum. He believed that economic facts were the outcome of human choices.

> It is by no means a consequence of the nature of man or of work, that two classes of citizens with opposing interests are necessary to carry out a piece of work. By this I mean the propertied class, who owns the [product of] accumulated work and who can relax; and the class of men who only have their vital force, and who sell their work. Their separation, their opposite interests, are the consequence of an artificial organization that we have given to human society. [This is] the work of man [thus] it is also subject to our censorship.[96]

Workers' needs depended upon what Sismondi called "government." The science of government, or "the administration of national wealth," should base its principles upon "man's physical needs." Each individual's needs were limited by their access to money and goods. The administrators of the national wealth must take responsibility to ensure that access. The fundamental end of government, for Sismondi, was to ensure "the participation of all citizens in the physical enjoyments that wealth represents."[97] Governments must set the conditions for the flourishing of needs, happiness and well-being.

The 1848 Revolution instituted, briefly, Sismondi's vision of a government of needs. In its first months, the Second Republic attempted to build an administration of well-being. The Luxembourg Commission, under the leadership of socialist Louis Blanc, was charged with rethinking the organization of work. National workshops were established to employ impoverished citizens. The *Republican Manual for Men and Citizens*, issued by the new Ministry of Public Instruction, promised that the state would guarantee its citizens' needs according to "the elementary principle of republican fraternity." Aid would be provided to the elderly, the ill, and the unable; for active workers, the republic would guarantee a "certain minimum income."[98]

Blanc envisioned an ideal society in which each citizen would "work according to his strength, and consume according to his needs." Those needs, he thought, would be "indicated by nature and assigned by morality." He believed, however, that a society governed by natural needs was impossible in the current state of France. Workers' true needs could not be fixed in a world driven by greed and "depraved tastes."[99] Instead, Blanc called for complete equality in wage rates, and a scientific "survey on the fate of workers."[100]

Hundreds of workers petitioned the National Assembly for a minimum wage. Historian Rémi Gossez found that among 616 petitions sent by trade corporations to the National Assembly in 1848, 472 called for wage increases. Petitioning workers called for a minimum wage set "according to their needs."[101] Several sought to establish a "jury composed of workers and bosses," independent mixed commissions who would investigate local needs and set a base wage in each trade.[102] During the short life of the Second Republic's Luxembourg Commission, Parisian workers turned it into a de facto guarantor of trade-based wage settlements.[103]

Many workers pushed to establish a moving measure, a wage tariff, which would respond to changing economic conditions. Petitions to the Assembly generally proposed precise figures for wage rates in specific localities and professions.[104] A delegation of roofers, for example, called for legal regulation of hours and pay: "The inevitable consequence of limiting the working day to twelve hours is to set a minimum wage in each industry, according to the needs of each arrondissement."[105] Several petitioners provided a budgetary list of "goods of prime necessity" to argue for a fixed minimum wage for their trade.[106] The *Salut du Peuple* published a plea in 1849 that the government should "as soon as possible, determine a minimum wage for each industry and each locality. . . . Where necessary, the wage should be proportional to the productive abilities of workers, the murderous influence of certain industries, etc. . . . For this, one should establish two or three categories of wage for each workshop or according to the type of industry."[107] Trade associations in 1848 promoted a statistical approach to needs and wage setting. They accounted for local, professional, and skill distinctions. Only fine-grained statistical measures, they argued, would reflect the changing share of individual workers in the benefits of their labor.

Conservatives like Gérando and Le Play sought human needs in nature and derived social organization from them. Schacherer, like his counterparts at

L'Atelier and petitioners to the 1848 Republic, blamed an unjust social organization for compressing workers' well-being. Needs were determined by historical circumstances, not by natural conditions of existence. Workers' needs were constrained by the power of the propertied classes to enjoy the fruits of workers' labor. This power was circumstantial and historical. The relative share of workers and capitalists depended on their political strength. Needs were not independent things in themselves; rather, they reflected the broader social division of power.

CHAPTER 5

Citizens

When a ship is in distress, taken by surprise by a flat wind or something else, we ration the crew in order to endure longer. Well! Paris is a vessel that carries the destiny of the Republic in its beams![1]

The Siege of Paris in the winter of 1870–71 presented exceptional circumstances of scarcity and constraint. Parisians fought for access to food under increasingly dire circumstances. Through the long siege winter, city residents confronted and debated the meaning of deprivation, suffering, and sacrifice. The dissident radical Auguste Blanqui, founder of the political club La Patrie en Danger, wrote that each citizen should receive a "minimum sufficient ration."[2] Liberal economist Gustave de Molinari, a paragon of parsimony, called for "everyone to limit his consumption to the strictly necessary."[3] Placards called for the expulsion of so-called useless mouths, the very poor and very rich. The question of needs was an urgent political issue, which continued well past the withdrawal of Prussian forces in 1871.

Both radical and liberal polemicists recognized the importance of food in the survival of Paris and therefore of France itself. "One thing is certain," wrote one journalist, "subsistence is one element of the defense of Paris, just as important as ammunition, war matériel and soldiers."[4] Another stressed the import of "munitions of the mouth."[5] The beginning of France's Third Republic was marked by this interweaving of subsistence rights and citizenship.[6]

Siege politics were all about needs. Under siege, the public debate over need turned around the stuff of daily survival: who legitimately needed meat and who needed bread. One of the strongest themes in siege politics was the distinction between "useful" and "useless" residents. The former described brave, active citizen-defenders who contributed whatever strength or resources they had to the cause of national "defense to the last."[7] The latter, populations outside the realm of the market, either excessively wealthy or excessively poor, were "useless mouths." The republican government defined citizens' needs according to their utility to the state.

The siege began when Napoleon III's blundering diplomacy led France into an ill-fated conflict with Bismarck's expansive Prussian state. The emperor himself was soon felled by his own war, taken prisoner with much of the French army at the battle of Sedan. The Second Empire collapsed in the absence of its leading personality, and the stage was set in September 1870 for a bloodless republican revolution. Legitimized by a clamoring crowd, opposition members of parliament and city officials proclaimed a Government of National Defense from their headquarters in the Paris city hall. The resulting regime change, however, did not put an end to the conflict. Prussian strategy turned increasingly toward containing the French civilian population. Bismarck planned to break the September 4 regime by starvation, weakening public spirit, and menacing its citizens' survival.[8] His troops surrounded the government's center of operations, the city of Paris. The Prussian siege transformed the urban experience: dimmed gaslights and candlelit streets, pedestrian circulation, and above all a constant struggle to provide for oneself.

The advancing Prussian army encountered an urban population mobilized for war to an unprecedented degree. Private letters and newspaper commentators repeated the same observation: "Paris is a great army camp." "Out of five people [on the streets of Paris], I am certain that one could count two mobile guards, one national guard, one combat soldier, and . . . one bourgeois."[9] By the end of September 1870, over three hundred thousand Parisian residents had signed up for the *garde nationale sédentaire,* a massive battalion of reserves formed to patrol the city ramparts and to assist in any potential counterattacks against the Prussians.[10] The amplitude of response to the government's call for volunteers, who were promised a daily stipend for themselves and their families, far surpassed expectations. "Paris besieged . . . is no longer a city, but an army camp; its population is an army, because all of its citizens today are soldiers."[11]

If all citizens were soldiers, they became employees of the state. To serve the new republic implied, in the view of many national-guard members, a state-guaranteed right to subsistence. Even before the final closure of the city walls in 1870, activists were calling for a subsistence administration consonant with the role of Parisian residents in the national defense. Attendees of a political gathering in mid-September at the radical club La Patrie en Danger heard one citizen "call for equality for all in the distribution of food. It is just, he said, that the defenders of the fatherland should receive aid, and that their families should be supported."[12] Public demands for a universal conscription coincided with claims that "rations should be distributed to all citizens in arms."[13] Under extreme circumstances of a besieged city at war, soldier-citizens and government officials debated the right to a minimum wage.

Radical propagandists claimed that the Parisian population under siege should be treated as a unified group of national defenders. The city had become a garrison, and residents assumed the role of soldier by fact or proxy. "For a city like ours, animated by such a high patriotic sentiment, the real peril is not the power of enemy arms, but the possibility of a weakening of public spirit due to the constant worry about subsistence needs, and the strain placed upon our innate egalitarian sentiment."[14] In October 1870, the activist Tapié brothers published a two-part pamphlet in favor of "a special regulation meant to ensure to all residents of Paris—indistinctly and equally—their daily food."[15] "Parallel to the right to perish, a right to life must be granted in common! . . . Today each resident of Paris could be required to contribute to its defense, in a manner proportional to the person's age, sex, and faculties, and [therefore] should be understood as a soldier with the right to be nourished by the collectivity whose interests he serves and whose liberty and rights he defends.[16]

Despite the Tapiés' rhetoric of egalitarianism, their republican right to life was neither universal nor equal. Their caveat concerning "age, sex, and faculties" was crucial. In fact, this exception served to exclude effectively much of the population from a right to subsistence. The new republic designed its rights and rations for active males. All others were relegated to charity or the category of "useless mouths." This, too, echoed the longer history of minimum wage debates. Male workers, defenders of the country, deserved a subsistence wage; women, children, and indigents were treated as charity cases. That structure would repeat itself in the welfare institutions of the Third Republic.

USELESS MOUTHS, GET OUT!

Despite many public calls for solidarity, most writers agreed that some residents should not qualify as members of the mobilized citizenry. "The nest is threatened," wrote one strident September call for revolutionary measures; "the weaklings must abandon it."[17] A strange conglomeration of people, in the early days of the siege, were labeled "useless mouths." This term served mainly to describe those unwilling or unable to defend the city, especially aristocrats and charity cases. It recurred frequently in political propaganda by both left- and right-wing writers.

The moniker "useless mouths" reveals a vision of population politics shared by many Parisians in 1870–71. This category differentiated between Parisian residents on the basis of citizenship and utility. If members of the active citizenry deserved subsistence by right, "useless mouths" defined those who had no such right to exist. Mouths without use to the nation could be banished from the polity or sacrificed to charity.

As the Prussian approach appeared increasingly inexorable in early September, some of the better-off Parisians decided to skip town. The departing "deserters" were met with scorn in the popular press. One engraving recounted:

> Many of the more fortunate of this world, those who easily carry off the fatherland on the soles of their shoes, have set course for the provinces or for foreign lands. . . . With the scared, the timid, the dandies and their friends, the dames have invaded the train stations to run to neutral countries. . . . The people laughed when they saw these useless mouths leaving town.[18]

Others encouraged such "useless mouths" to exit the scene before provisioning became tight. In a public lecture during the siege, Professor Edme Bourgoin of the School of Pharmacy warned that Parisians would suffer from deprivation of fibrin if their meat consumption declined any further. He suggested that Parisians make horse meat a regular meal and that "useless mouths" be sent out of town. Bourgoin defined useless mouths as "every individual who is inevitably a consumer but who is not a producer at the same time." The most useless of all had already been sent away: that "veritable mouth of Gargantua" Emperor Napoleon III.[19]

As high-heeled ladies sped far away on the train tracks, the roads outside of Paris were packed with a different variety of nonconsumer. Louis Denormandie, mayor of the Eighth Arrondissement, recalled: "They said that the useless mouths—this was the accepted expression—had already left Paris in September 1870 and that the municipal administration therefore had fewer mouths to feed. This is an error. . . . Paris saw as many mouths enter its walls as had left them, mouths that were unfortunately just as useless, but that humanity obliged us to receive."[20] Displaced inhabitants of the entire region surrounding Paris—several thousand of them—came on foot, with carts and horses, seeking shelter in the fortified city. At the same time, drastically increased hardship and a reform of the Assistance Publique multiplied by a factor of four the number of Parisians on the rolls for poor aid in 1870 and 1871.[21] Jean-Jules Clamageran,

FIGURE 6. Daniel Urrabieta Ortiz y Vierge (1851–1904). La queue aux cantines nationales (novembre 1870). Pen, graphite. Paris, Musée Carnavalet.

adjunct to the central mayor of Paris, estimated a total of at least four hundred thousand needy people in the city.[22]

Some Blanquist proponents of rationing sought to turn all of Paris into a community of citizen-soldiers. As soon as the republic deposed the Second Empire, radicals campaigned to eliminate the previous regime's charitable institutions. Leftist Eugène Razoua and his National Guard battalion marched to the seat of the Government of National Defense and demanded the abolition of the imperial Assistance Publique.[23] Polemicists at La Patrie en Danger argued that the poor should be included in the ranks of the city's defenders. "We know that there are soup kitchens and other such institutions. But these establishments . . . are mostly run by nuns, who distribute food with an air of importance and protection that can cause the wife of an honest worker only to suffer. . . . We are not asking for alms of any kind. We are claiming our right to live while fighting. If they do not give it to us, we will be forced to take it."[24] If the new republican citizens refused to be fed by charity, they did not indicate the fate of those who did not or could not "live while fighting."

Parisian municipal rationing reinforced the divide between "citizens" and "useless mouths."[25] The population of food seekers was effectively partitioned between municipal soup kitchens and retail stores. Ration cards carried meaning only for those who could pay shopkeepers for the quantities specified upon them. Meat rationing in the Parisian autumn of 1870 raised only a minor hubbub. Bread rationing, by contrast, set off a strong and violent uprising.

The national government soon recognized that its first responsibility to its citizens was to feed them. As Mayor Jules Ferry put it, "The government has a duty to look after the people's subsistence."[26] Amidst the grave shortages induced by the Prussian siege, this would be no easy matter. The advancing enemy blocked the circulation of rail traffic, and the city was forced to make do for the duration with existing stocks. Even political cant acknowledged officials' limitations in that regard: "We should not deceive ourselves: whatever we do, a crisis like the one we are in will upset certain interests, impose privations and suffering; the administration . . . will not refuse any means to ensure the people's subsistence for as long as possible, but it is counting on the citizens' patriotism to help in the measures that it believes necessary."[27]

Rationing first appeared in besieged Paris at a local, haphazard, and nongovernmental level. During the siege's first month, neighborhood sections of the National Guard implemented their own provisioning order. One outraged commentator reported that "some national guards are interceding in the relationship between clients and butchers, not to maintain order but to appoint themselves as distributors of meat. These national guards force the butchers to cut up their products and to sell portions of equal weight to each customer in line; naturally this distribution is the least egalitarian of all, since some who need many pounds are left needlessly with the measured portion, while other housewives, who want or can only afford one pound, receive two or more pounds."[28] In addition, guard members across the city formed small local *conseils de familles*, which catalogued needy residents and distributed poor aid in their neighborhoods.[29] Local arrondissement mayors improvised distribution systems as best they could, according to their particular economic predilections.

Rationing held vastly different meanings for the many consumers who participated in it, and it imbued consumption patterns with political significance. Besieged Parisians defined themselves in and through consumption: where and how they shopped, ease and proximity of access to markets, the goods they obtained. The difficult economy of basic goods during the siege provided a new set of terms for creating social distinction. Some experienced the siege as an inconvenience, others as a moment of solidarity in suffering. Some succumbed anonymously to hunger-related disease.[30] Stéphane Rials estimates that between September 1870 and February 1871, Paris had an excess mortality of forty-two thousand compared with one year earlier.[31]

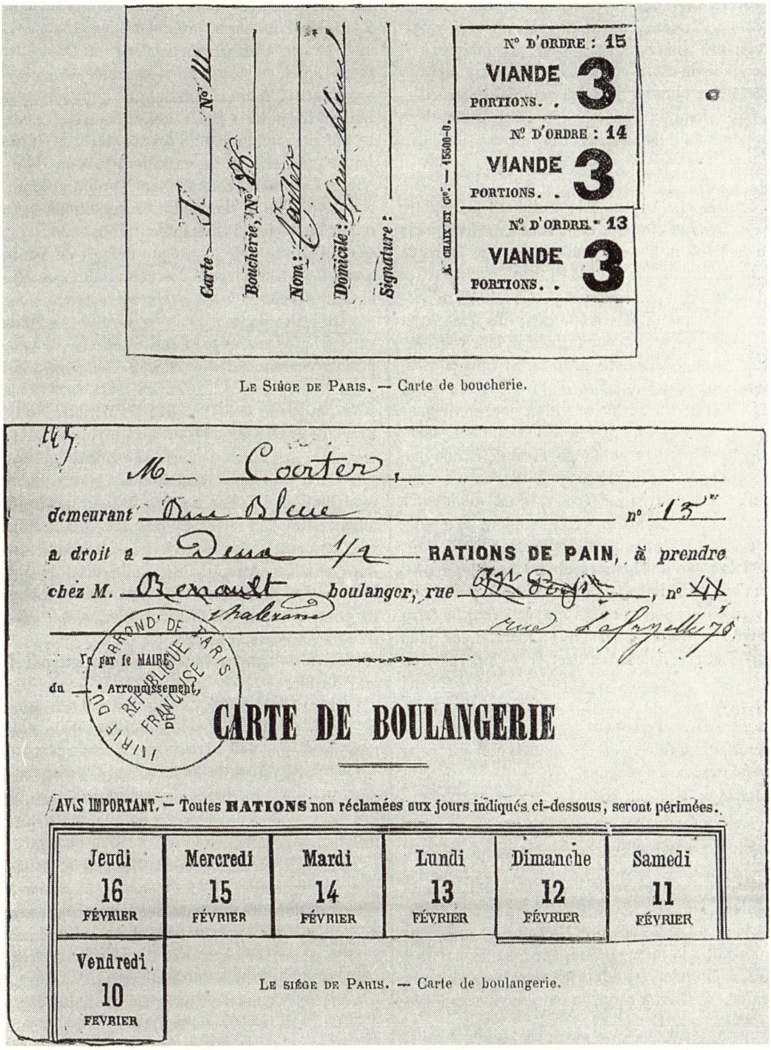

FIGURE 7. Paris, France: War of 1870. Siege of Paris. Rationing card of bread. © Roger-Viollet / The Image Works.

Though chaotic and spotty, a technical apparatus for rationing was fully in place by the end of 1870. Local tabulations, then a full-scale city census, created a demographic database of the city. The census distinguished between active and needy populations, and it recorded the size and age range of each family. Local mayors registered residents, sometimes proceeding with home visits to count their families and verify their financial situation.[32] Officials intervened actively in market exchange. Basic goods were removed from the free market by requisition and purchase. Most residents were issued ration cards for meat, followed by a few other essential products. These cards were printed and administered by the mayors of individual arrondissements. Each family received a card that measured provisions daily or weekly.

Between them, Parisian municipal canteens and the Assistance Publique's "economic kitchens" distributed thousands of charitable meals daily. Mayor Etienne Arago estimated that the *fourneaux économiques* provided over two hundred thousand meals per day.[33] Diners in these establishments paid five centimes per serving of bouillon, a tiny slice of meat, a couple of vegetables or 125 grams of bread.[34] These measures provided a thin layer of aid in a city where up to four hundred thousand people, around a fifth of the population, lacked the means to care for themselves.[35] The inadequacy of governmental support was made far worse by exposure to bitter cold, lack of fuel, and disease.

Liberal economist Gustave de Molinari combated government rationing with vigor and force in numerous editorials to the *Journal des débats*. Molinari, a protégé of Frédéric Bastiat, had become an influential voice in French political economy through his position at the *Journal des économistes*. Molinari refused any state intervention in Parisians' subsistence, either by ration cards or by daily stipends to national guardsmen. He believed that the market itself would restrict consumption most efficiently. He reminded his readers of "the disastrous experience of the first Revolution," which showed that when the government interferes with exchanges between sellers and consumers, "it aggravates the problem rather than curing it."[36]

Instead of regulating prices and rations, Molinari proposed, municipal authorities should distribute food aid in kind. Authorities must guarantee that the food they provided was only for consumption and could not be resold.[37] Molinari quoted at great length the infamous liberal editorialist of the Irish famine, Charles Edward Trevelyan. Trevelyan is known for inventing means to wean the Irish masses off the rolls of English government assistance. Molinari particularly admired the formula whereby English administrators substituted ground flour, which aid recipients could potentially stock and sell, with a precooked mixture of flour and vegetables that soured quickly and "had no market value."[38]

Jacques Siegfried, president of the central government's Subsistence Commission, proposed a three-tiered system of municipal distribution: at one end, the most basic necessities would be given as charity to those stripped of their livelihood. At the other, luxury items would go to the highest bidders at auction. In between these two sets of useless mouths, a third portion would be sold in retail shops.[39] This arrangement essentially came to pass. Private charitable societies held auctions that purveyed rare items like cheese and leather boots to government ministers and wealthy businessmen.[40] Ration cards went to all residents who had the means to shop. Finally, soup kitchens distributed variable amounts of aid to the dispossessed.

MEAT OR BREAD

> [Either] allow the defenders of Paris to die of hunger, or take revolutionary measures to feed them. Reflect upon this: a hungry man is beyond the law; he has the right to take bread wherever it is, because he hasn't got any.[41]

In the mouths of Parisian consumers, particularly of the poorer classes, bread formed the anchor, the staple, and the overwhelming preponderance of nourishment. The three-to-one proportion of bread to other edibles in the nineteenth-century soldier's ration conveys some measure of its significance. Steven Kaplan has thoroughly demonstrated the detailed exigencies that French urban residents have applied to bread since the eighteenth century. Its weight, color, and texture all carried grave cultural meaning. White bread, a symbol of artisanal refinement, was the only variety accepted by even the poorest of Parisians.[42]

The central government responded to the Prussians' arrival in September 1870 by immediately bolstering the price controls already in effect on bread.[43] When flour stocks grew thin, the municipal government chose to affront shoppers' taste rather than to resort to rationing. Mayor Jules Ferry dodged Blanquists' and national guardsmen's lobbying for universal rations. "Instead of rationing, we changed the quality of bread. It was, I admit, detestable bread. I will carry the responsibility for it to my grave."[44] The central government's Flour Mill Service, directed by social statistician Emile Cheysson, gradually reduced the amount of flour in the prescribed composition of bread. By the end of the siege, wheat flour entered into only 25 percent of each loaf, the rest of the bread being filled with rye, barley, oats, rice, and potato starch.[45] Souvenir posters and knickknacks, all the rage after the siege's end, often held a small piece of "siege bread" as testimony to its foul quality and the suffering of those who ate it.[46]

Rumors and public violence almost exclusively targeted bread and bakers. "Paris is peopled with alarmists," wrote one perturbed commentator, "who will tell you in lowered voices that the administration does not have as much

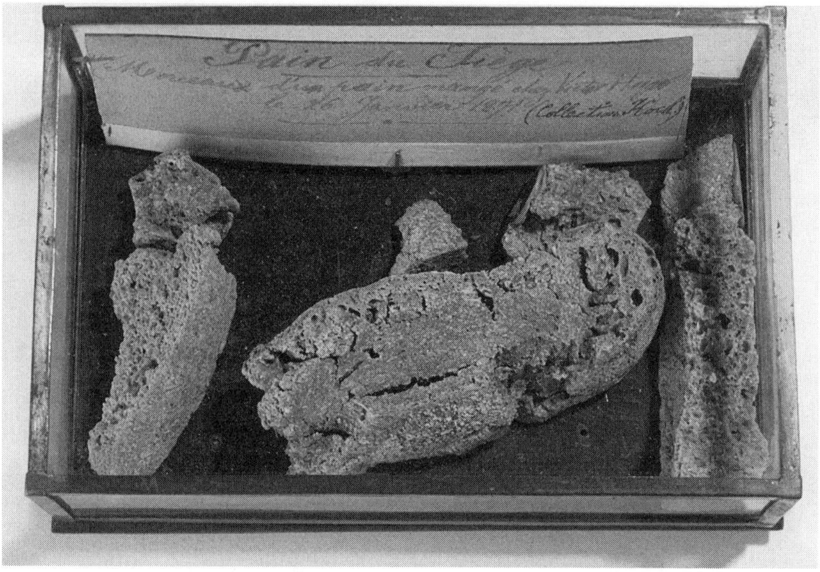

FIGURE 8. Bread dating to the Siege of Paris in 1871. © PJoffre / Maisons de Victor Hugo / Roger-Viollet / The Image Works.

wheat and flour as people believe they do."[47] Markets and shops were regularly the scenes for "seditious cries and furious shrieking [*glapissements*]."[48] A violent tremor erupted in mid-December 1870 over rumors of imminent bread rationing. The prefect of police at the time recalled the events that followed:

> Word of a probable [government] decision on rationing had spread . . . and the commotion became universal. The bakeries, instantly surrounded, emptied of bread and flour, were besieged by newcomers full of suspicion and accusatory rumors, and closed. . . . The [police] reports of December 9 and 10 indicated preparations for a demonstration of people from the faubourgs. The rally would have women and children at its head: plans were to carry off bags of flour and merchandise from the bakeries in the center of Paris, even in exchange for money: the pillagers would charge themselves a tariff. These threats were realized.[49]

In the popular neighborhoods of Batignolles, Belleville, Ménilmontant, and Gros Caillou, crowds gathered. "National guards seized the bread in a delivery cart and sold it to passers-by on behalf of the baker."[50] The uprising only cooled when the Government of National Defense and the mayor of Paris issued an official denial—"bread consumption will not be rationed"[51]—which the prefect of police hastened to print and paste on the city walls.[52]

The government waited until the final hour to renege on its promise. On January 13, 1871, bakers were obliged to sell only to their regular clientele or to those who could present a rationing card attesting to a local residence. A few days later bread itself was rationed throughout Paris at 300 grams daily for adults and 150 grams for children under six.[53] This quantity represented less than half of the soldier's ration, and less than three-quarters of average daily consumption during normal times.[54] Those who depended upon bread for their sustenance took a serious cut in caloric intake.

Bread rationing sundered the Parisian population into sharply distinct classes of consumers. Bourgeois shoppers felt the pain of bread rationing "even more in the soul than in the body." "[It] was very painful, because of the general weakness, the almost absolute lack of other foods, the cold, the detestable quality of bread, and above all because of its significance. . . . It seemed like the necessary prologue to surrender."[55] Pastor Vesson viewed the situation more philosophically: "This measure had been demanded by public opinion for a long time, and was welcomed with the greatest resignation. Every one of us said to himself, rubbing his hands—and tightening his stomach—that this would prolong Paris's resistance for a few more weeks."[56] By this late date even moderately wealthy inhabitants of the city certainly must have suffered physically and mentally from hunger, cold, and discomfort. However, to this group bread rationing represented only one more difficulty to surmount for the national cause.

For lower-class Parisians, the measure bore catastrophe. At the Favié political club in Belleville, residents expressed their panic and hatred.

> Yesterday evening: the Favié club was a hive of activity. [The people there] spoke animatedly about the bread rationing that had been instituted in the arrondissement that same morning, and women, in particular, distinguished themselves with the vehemence of their complaints. The folks in the Hôtel de Ville have it easy, the people said; they and the rich who support them eat as much meat as they want. . . . They are not suffering from bread rationing, and it is easy for them to be patriotic and "defend to the last" at the expense of our stomachs. But we, who have nothing more than bread to eat, can we live on a livre or even 400 grams of bread per day? . . . We cannot be the only ones to bear the torments of hunger.[57]

The republican spirit of privation had no place at the Favié club in the wake of bread rationing. For these people, the ration appeared to represent a limit lower than death itself. "They ration us, we who live on bread; we must seize the goods of those who live on preserves and fine foods. Before we die, we will go visit their cellars and we will have a few words with their hams. After that, we will die all together, and since they did not want communal life with us, we will have it with them in death."[58] The drama that unfolded around bread in January 1871 revealed differences at the core of republican rights and needs. Rations and the right to subsistence grounded an egalitarian vision of citizenship in the new Third Republic. At the same time, this very model of citizenship excluded entire social groups of women, children, and indigents from a "right to exist."

CHAPTER 6

Vital Wages

In April 1887, the newly elected radical-socialist Paris city council passed a rule on wages in public works construction. "The City of Paris's official [wage] series will be revised annually," the council decreed, "such that the wage rate will always remain in relation to the price of subsistence and the general conditions of workers' existence."[1] For the first time, consumer budgets appeared as a ground for official wage regulations. The outraged deputy Auguste Bouge, enemy of all forms of "collectivism," angrily denounced this turn to a need-based wage standard. With a municipal minimum wage in place, he grumbled, it would be "useless to evaluate a worker's work, all one needs is to estimate how much he needs to live."[2] This tight association between needs and wages is precisely what workers' organizations under the Third Republic fought for with increasing urgency. The modern social order, the politics of wages and welfare were articulated in a language of needs.

Wage debates in twentieth-century Europe were dominated by a new regulating principle: the *minimum vital*, "living wage," or *Existenzminimum*. Below I analyze the political-scientific content of the "vital" or "living" component of the living wage. The figure of the vital minimum traveled from classical political economy to physical energetics, to Marxism and social Catholicism, to sociology, to Fordist wage regulation and welfare. The vital minimum performed multiple rhetorical and political functions, often contradictory. It could refer to a physiological threshold, grounded in chemical-thermodynamic studies of individual male model organisms (workers). Or the vital minimum could describe a collective cultural and sociological norm, subject to change over time. Welfare politics invoked both physiology and sociology via the vital minimum. Strikingly, the French welfare state first mobilized the vital minimum as an individual minimum wage. Later it changed sites and reappeared as the logic behind a collective family allowance. The vital minimum elided scientific objectivity, physiological constants, and class politics.

The term "vital minimum" first appeared in a popular pamphlet written by Jean Jaurès at the turn of the century, on the eve of founding the French Socialist Party.[3] Many French and German socialists postulated that capitalism tended to push workers' wages below their minimum subsistence needs, biologically and historically determined. This was the "iron law of wages," or as Jaurès put it, the "limit of the exploited classes' vital tolerance."[4] As the expression "vital minimum" spread into workers' journals and union halls, it bolstered the idea that wages had a bottom limit, a floor that should not be broken.

The vital minimum, shorn of its Marxist associations, was a ubiquitous trope in twentieth-century France; the term is still in common usage today.[5] Its power derives in large part from its polysemy. Marxists defined a vital minimum limit to wages, as a critique of capitalism; many trade unionists took it as a justification for minimum wage legislation. In the early twentieth century some social Catholics picked up the term "vital wage" and imbued it with a moral meaning. During the First World War, the French government set a "vital wage" standard for workers in war industries. The wartime administration convinced both workers and employers to accept a mandatory vital wage by invoking its multiple meanings. By the end of the war the "vital wage" referred primarily to family allowances, wage supplements destined for the upkeep of women and children. The "vital minimum" became a cornerstone of social policy under the Vichy regime and the early postwar Fourth Republic.

"Vital minimum" suggests a physiological standard, grounded in medical authority. Vital, in its literal meaning, signifies "essential for life."[6] In the eighteenth century "vital principle" defined a life force unique to organic beings. Vitalists viewed living beings as holistic and endlessly variable entities, which could not be reduced to their physical and chemical components.[7] A different kind of vitalism emerged in French physiology around the turn of the century, this time derived from energetics. Vital force meant energy, and the vital minimum measured calories and protein. These measures filtered into political debates. The Marxist iron law of wages generally referred to a physiological threshold of survival. Some conservatives echoed this model as they tried to distinguish between pernicious, "false" needs and "real," valid, physiological needs. The vital minimum seemed to imply an inexorable, static natural limit.

Despite this, and despite the role of physician-legislators in shaping the early Third Republic, doctors and chemists played surprisingly little role in wage debates during this period. In the wake of the Prussian defeat of 1871, doctors sought to strengthen the nation's population, its quality and quantity.[8] Physicians, outraged by the weakness of the French nation and its ineffectual sanitary administration, took to politics. There were more doctors than any other profession, except lawyers, in the Third Republic parliament.[9] They enacted a politics of population: pronatalist programs to raise the nation's birthrate, neonatal care to preserve infants from death, salubrious housing and vaccinations to prevent infectious disease. Doctor-legislators instituted a pillar of the modern French welfare, social insurance for illness and disability. This was the age of hygienic solidarism, shaped by fears of contagion and heredity.[10] To guarantee the nation's health meant to intervene in the private lives and behavior of all French citizens.[11] Hygienists of the Third Republic rejected their predecessors' liberal politics and sought social, proactive solutions to maintain the nation's health. Their sphere of action, however, remained similar. They focused on the home, the family, and the street, largely eschewing questions of class, labor, and wages.[12] Wage work, in the hygienic view, lay outside of the collective sphere and the realm of social legislation. The vital minimum was elaborated not by doctors but by statisticians, melding natural science and sociology.

SOCIALISM, STATISTICS, AND THE IRON LAW

Jules Guesde, founder of the Parti Ouvrier Français in 1882, warned workers that their campaigns for higher wages were doomed to failure. Wages could never escape the inexorable logic of capitalism: "The limitation of wages to a minimum of subsistence necessary for the worker to live and reproduce himself is the natural and inevitable result of an order in which the worker does not exist by himself and for himself any more than the . . . locomotive or horse."[13]

For Guesde, minimum wage demands served only to demonstrate that under a capitalist regime workers would never receive decent pay. No wage law that guaranteed decent living conditions would ever pass through the parliament. "For us, [this impossibility represents] . . . a condemnation of capitalist society, obliged to confess its powerlessness to guarantee even a minimum of existence to those whose work alone allows all of us to exist."[14] Guesde, like his mentor Karl Marx, discounted the influence of union activity or legislation on the minimum limits of wages. Strikes and public protests could only slow the descent of the lowest wages even farther into misery.

This limit, Guesde specified, represented the basest "organic needs" that would render a worker capable of a given industrial task and keep him from death.[15] Marx, at certain points in his writings, put forth this dire prescription. "The very nature of the wage relation, demographic pressures and the demand for ever-increasing surpluses," Marx wrote in *Capital*, "forced workers to hover at the very frontier of their physical survival." Marx repeated a story told by German chemist Justus von Liebig, that miners in South America were forced to eat beans with their bread so that they would have sufficient strength for their work.[16] Inadvertently, Guesde slyly remarked, classical economists had bequeathed "a weapon—what a weapon!—to socialism" with their principle of a "necessary wage." Guesde took this as an admission that workers could only ever hope to receive the smallest possible subsistence.[17]

The theory that bound remuneration to a strict physiological minimum became known in popular literature as the "iron law of wages." German socialist Ferdinand Lassalle coined the "iron law" in an 1863 pamphlet on "the workers' question."[18] Lassalle quoted David Ricardo and John Stuart Mill to argue that wages would inevitably fall to the "minimum of needs for existence [*minimum Existenzbedürfnisse*]."[19] The only path out of wage oppression, in Lassalle's view, led to revolution. The German *Existenzminimum* and the French *minimum vital* initially referred to a purely physiological measure.

Socialists of the late nineteenth century read and integrated nutrition research into their theories of "organic needs" and "minimum existence." Karl Marx had cribbed extensively from Liebig's physiological and agronomic works.[20] Later socialists drew from the work of Liebig's successors, physiologists Carl Voit and Max Rubner. Nutrition science figured in an early version of the standards of living debate. Conservatives argued that workers' lives had improved greatly under industrialization, as evidenced by increased meat consumption.[21] Socialists like Guesde responded that industrial work required greater physical force and therefore a higher, protein-rich minimum diet.[22]

In the late nineteenth century, German scientists reshaped the field of nutrition. Voit and Rubner applied the new thermodynamics to the human body. Rubner posited that all life processes were energetic; the body's chemical inputs and outputs mattered far less than the intake and expenditure of energy. Rubner adapted a physical measure of energy to physiology and invented the calorie. As long as it contained the same amount of energy, he postulated, any food could substitute for another.[23] Instead of measuring the carbon and nitrogen necessary for a "maintenance ration," along the lines of Dumas and Boussingault's model, nutrition scientists sought to determine minimal caloric and protein requirements.

French physiologists translated the Germans' language of energy into one of "vital functioning."[24] "The first and foremost goal of an organic body [*être organisé*] . . . is the normal exercise of its vital functions."[25] In the French version of the energetic human body, food provided "vital energy."[26] It cannot be coincidence, then, that French Socialists employed the term *minimum vital* in the same sense as the German *Existenzminimum*.

Politically, Lassalle's notion of *Existenzminimum* in Germany followed a parallel trajectory to the French *minimum vital*. Whereas Lassalle posited it as a critique of capitalism, other Germans used *Existenzminimum* to define a proactive minimum wage limit.[27] A similar shift took place in France. Guesde's mechanical view of wages, falling inexorably toward minimum subsistence, was marginal even within his own party. Jean Jaurès, co-founder with Guesde of the Parti ouvrier, openly mocked the belief that falling wages would inevitably lead to the overthrow of modern capitalism. Instead Jaurès directed workers toward "progressive organization" and "social transformation."[28] Socialist workers pushed heavily for a legal minimum wage.

Labor activists in the late nineteenth century turned to social statistics, not physiology, to bolster their wage demands. The Federation of Socialist Workers, later known as the Parti Ouvrier Français, voted a "legal minimum wage" in 1879 as part of its political platform. The legal wage level would be "determined each year on the basis of the price of goods [*denrées*], by a workers' statistical commission."[29] Workers' commissions, given the authority to define a minimum level of needs, would "carry manual laborers to the highest possible level."[30] "In order to obtain a minimum for wages," the unions stipulated, "it is urgent that each union gather statistics of daily work, wages, work stoppages, food supplies, rents, etc., and send them to their federal committee, according to their region."[31] Union platforms called for a network of worker-statisticians, organized by trade and region. They claimed workers' power to control the measuring apparatus and establish a minimum wage.

These efforts formed part of an international movement for working-class statistics. When the International Working Men's Association held its founding congress in 1866, its first substantive resolution pledged to gather comparative statistics. Members would send in data on wages, the state of industry, and "the respective situation of the working classes in all countries." Only then would workers be able to unite and "act with full knowledge of the facts."[32] Some at the congress anticipated that statistics would allow workers to arbitrate differences in wage rates across various countries. English workers, in particular, hoped that public knowledge of common wage rates would discourage continental workers from undercutting locals in the British labor market.

Many French workers likewise hoped that a fixed minimum wage would drive out cut-rate foreign and women workers. Early minimum wage legislation was explicitly designed to "protect" industries from cheap female labor.[33]

Anarcho-syndicalists anticipated that statistics on labor supply and demand in different industries would allow workers to self-organize.[34] Workers could monitor economic conditions and move to areas in demand, without the interference of employers or the state. This was the principle behind the Bourses du Travail. Another group put workers' statistics in the tradition of *cahiers de doléance*, accusatory evidence against capitalist exploiters.[35] These were closer to the monographic tradition of the tableau, a picturesque overview of poor conditions illustrated by arithmetic tables. Ferdinand Pelloutier and the Bonneff brothers are the best-known French exemplars of this tradition.[36] Still others viewed statistics as an instrument to force a legal minimum wage.

When French trade unions established a common program at their national congress in 1894, they called for a minimum wage that "guarantees the strictly necessary needs of material life."[37] French wallpaper workers declared that "the rising cost of basic necessities [*objets de première nécessité*] should have as its correlate a rise in wages."[38] The Federation of Construction Workers in 1881 demanded a wage proportional to "the price of rents and foodstuffs in the locality."[39] In 1883 the Professional Society of Parisian Mechanics set out to "determine and maintain a minimum wage rate . . . proportional to the price of things necessary for existence."[40] Consumer budgets served as support for the legitimacy of wage claims.

The financial, industrial, and agricultural crises of 1873–96 lent urgency to workers' wage demands. During the economic expansion of the mid-nineteenth century, wages and prices had risen roughly in tandem. The 1870s depression ended this trend. The nominal price of wages came uncoupled from the cost of living.[41] Wages stagnated and urban housing costs soared, even as consumer purchasing power increased. The prevailing sense of instability was exacerbated by a tight labor market. Employment creation slowed to a stop for thirty years and layoffs loomed.[42] In the first decades of the twentieth century, nominal and real wages split in another direction as the cost of living soared with inflation.

As an urban mass market developed in the late nineteenth century, the line between basic needs and luxuries grew blurry. Urban consumers purchased Argentine beef, American flour, and cheap textiles, while industrial exports entered a downward spiral. France's domestic market was submerged in a globalized economy. The creation of consumer credit in the 1870s accelerated this trend.[43] Limited forms of credit (which still remained rare in France well into the twentieth century) brought items like sewing machines into the homes of working women.[44] Under these conditions, workers were constantly forced to reassess the real value of their wages and the horizon of their aspirations. Workers measured their wages not by their nominal value (which remained flat) but by the lifestyle that they could afford. They refused to rely on official numbers to tell them how much they needed to live and set out to measure minimum needs themselves.

The depression of the 1870s put fatal pressure on informal systems of wage regulation. The City of Paris had collected statistics on pay rates in the building industry since the 1830s. Like similar local tariffs for bread, these figures had gained moral, if not legal, authority.[45] When the depression hit, most Parisian contractors withdrew altogether from the city's wage commissions.[46] Only at this juncture, following the complete breakdown of semi-official and negotiated wage standards, did the municipal council propose a legal minimum based on a consumer budget.

Less than a year after its passage, the national administrative high court voided the Parisian minimum wage as an obstruction to the freedom of contract. The high court's forceful reaction did not prevent socialist city officials elsewhere from passing a series of similar decrees, all vetoed by regional prefects.[47] The resulting stalemate opened a decades-long struggle to enact national legislation on wages. A universal minimum wage did not appear in France until the Second World War.

What kinds of social aggregates did workers' statistics create? Statistical entities are objects of knowledge and of social action. To classify a population is to set out what can be done with it.[48] On one level statistics may operate as a collection of individual data points. Detailed data on wages, consumer budgets, and working conditions often appear in monographic form. Statistics on working and living conditions of workers in specific professions or regions open possibilities for individual action. A worker in Marseilles will know what wage to demand should he choose to relocate to Lyons. Some, like the Paris Commission on Work, called on statistics to arbitrate supply and demand. The commission's stoneworker's budget offered an individualist view of the wage exchange as a mediation between an individual's needs and an employer's means. This exchange was understood to be the act of an active male citizen, to the exclusion of women and foreigners.

Other views of workers' statistics set forth a collective social body, the working class. When workers offered statistics in the form of an accusatory *cahier de doléance*, they portrayed workers as a collective, the oppressed class. Political movements for a legal minimum wage promoted a different view of the social collective. An individual worker's consumer budget served as a typical model, in the sense of Le Play's representative samples. Budgetary studies like those of the Paris Commission assumed cohesion of social groups within professions and regions such that any one member could stand in for the whole.

The Paris Commission on Work presented the municipal council in 1896 with a model budget for its wage legislation: a stonecutter's expenses, based on a yearly employment of 255 ten-hour days.

> Daily expenses.
>
> The morning journey to work, on foot, imposes the need to have a glass of wine and a sou of bread along the way, equaling . . . . . . 20 centimes
> 9 am, breakfast . . . . . . 1.5 fr
> Noon, a snack . . . . . . 20 c
> 2 pm, tea . . . . . . 1 fr
> 5 pm, snack . . . . . . 20 c
> 6 pm, supper . . . . . . 1.25 fr
> Tooling expenses . . . . . . 50 c
> Return on the omnibus . . . . . . 15 c
> Total . . . . . . 5 francs
> Expenses during 255 working days 1275
> During the 110 days without employment,
> expenses are less by . . . . . . 412.5 francs
> Rent, at minimum . . . . . . 250 fr
> Clothing, shoes, heating, pressing, minimum estimation . . . . . . 250 fr
> Total expenses . . . . . . 2185.5 francs
>
> Wages are at 1785; the deficit is thus 402.5. This can only be made up by self-privation. Such is the situation of a single man; it is greatly aggravated for a man at the head of a household [*père de famille*] . . . . . .[49]

At the root of this new pay standard we find the type of budget well developed by social economists in the first half of the nineteenth century. Yet the budget's content reveals an intensified focus on the individual male worker and his expenses. Whereas the commission worried in detail about the worker's pedestrian commute and his need for a tipple along the way, it did not bother to itemize living expenses beyond clothing and heating. Most importantly, family needs disappeared from this minimum budget, only to appear as an admonitory mention at its end. This is a need-based minimum wage, but one in which needs have been redefined almost exclusively as a male work ration, geared for pure production.

One stonecutter stood as representative of all building workers in Paris—understood as single men. His representativity depended upon the expertise of the person who selected him. In that sense this budget functioned exactly in the manner of Le Play's monograph: it was valid because the data collector shared his subject's world and knew it by experience. Worker-statisticians explicitly raised the question of expertise in their own data gathering.

Most workers' statistics were designed to represent the working class as a cohesive entity, which existed above and beyond the lives of its individual components. This holistic vision of social groups echoed contemporary social scientists. Auguste Comte, Adolphe Quêtelet, and Emile Durkheim set out "the framework of a macrosociology in which the 'social' had a reality external and superior to individuals."[50] At the same time that French trade unions undertook their statistical project, sociologist Emile Durkheim posited "the existence of a collective type external to individuals."[51] He drew from Adolphe Quêtelet's "average man," which turned the average of accumulated individual measures into a scientific and a social entity. The average man was imbued with his own identity. For Quêtelet and his contemporaries, the average man had significance separate from the individuals whose measures contributed to define him. Durkheim defined social collectives not by averages but by an organic view of social cohesion and normativity.

The most forceful promoters of statistics within the workers' movement were the positivists. Inheritors of Auguste Comte, working-class positivists saw society as an organic whole composed of functional parts. Members of the Circle of Proletarian Positivists, founded by the executor of Comte's estate, were key figures in promoting statistics to solve the problems of labor. Isidore Finance, housepainter by profession and president of the Circle of Proletarian Positivists, actively promoted scientific solutions to labor and wages in workers' education programs and propaganda. "The proletariat," claimed Finance, "must familiarize himself with the notions of sociology and a social, scientific, proven moral order, which fixes and determines the relationships between all social classes."[52] Just as one must employ biology to understand the needs and functions of individual men, one must use sociology to determine society's collective norms. "Through sociology, we penetrate into the organism of each social group, from the humblest—the family—to the largest—humanity."[53]

Finance and his successor at the Circle of Proletarian Positivists, Auguste Keufer, played a key role in establishing the first official labor statistics in the 1890s.

State agencies, swayed by a coalition of statistics activists and labor activists, began in the late nineteenth century to systematically gather data on wages, family budgets, and the cost of living. In 1891 the Ministry of Commerce established the French Labor Office; this agency took control of the national statistics bureau that had been founded in 1801 by Lucien Bonaparte (see chap. 3). A group of intellectuals and reformist socialists coalesced around this program, including mining engineer Arthur Fontaine, worker-positivists Isidore Finance and Auguste Keufer, Socialists Alexandre Millerand and Albert Thomas, and professors Lucien Herr and François Simiand of the Ecole Normale Supérieure. The Labor Office "constituted the first scientific-administrative institute endowed with significant means. An ancestor of the Ministry of Labor, created in 1906, it combined services for conducting surveys and research into wages, unemployment, and workers' living conditions with more managerial services."[54] Statisticians combined Le Playist monographic methods and positivist statistical arithmetic.[55] The Labor Office succeeded in melding socialist and conservative ideologies, Le Playist and positivist methods. In part this was because most of its members shared a collective view of the class question.[56]

The tradition of trade union statistics fed into the best-known French sociology of consumption by sociologist Maurice Halbwachs. Halbwachs gathered data through a long active network of trade union statisticians.[57] He served in Socialist Albert Thomas's cabinet during the First World War, charged with organizing the production of munitions and military equipment. Thomas's ministry was a meeting point for leftist statisticians, economists, and sociologists seeking to establish quantitative data for social organization.[58] Halbwachs circulated in this space of trade union and statistical activism.

Halbwachs held a holistic view of social class as a cohesive social entity. He believed that consumer needs and choices revealed information about social norms and classes. The prevailing notion among economists was that purely personal preferences could explain consumer behavior. Marginalist economists like Carl Menger and Léon Walras assumed that people define their own needs by utility: they maximize their pleasure and minimize pain. Halbwachs countered that needs were always shaped by history, labor conditions, and, above all, class. Individual consumers had to be studied in their social context, especially their class position.[59] He considered his task to "re-situate social economy in sociology and history."[60]

By his second book-length study on need in 1933, Halbwachs grew uncertain that modern consumption could be measured against any kind of set standard.[61] The level of comfort achieved in the 1920s by Fordist American workers mystified and troubled many French social thinkers.[62] Class distinctions seemed to melt away when faced with mechanically produced mass consumption, consumer credit, and travel agencies.[63] On the basis of an extensive survey of family budgets from France, Germany, and the United States, Halbwachs concluded that social groups set "levels of living" [*niveaux de vie*] each according to its own particular values.[64] He concluded that no fixed order of needs was possible; all the sociologist could do was to track changing needs through regular surveys. "Would it not be scabrous," he reflected, "to refuse an individual his own scale of pleasures?"[65]

THE FEVER OF NEEDS

Emile Cheysson was typical of conservative social scientists on late nineteenth-century state commissions like the Labor Office. Cheysson was a self-designated "social engineer" and an omnipresent promoter of neopaternalist reforms. His biography reads as a summary of French social thought and policy under the early Third Republic.[66] A partial list of his activities paints a picture of great para-political activity in this period. He served as president of the Paris Statistical Society, the Society for Hygiene and Public Medicine, the National Agricultural Society, and the National League against Alcoholism, among many others. He sat on the national High Council on Work when it debated minimum wage regulations in 1897.[67] He was Frédéric Le Play's most devoted student and inherited the directorship of Le Play's institute for social reform and the Musée Social.[68] In the early 1870s Cheysson served as manager of the Le Creusot iron and steel works complex; this became an "industrial laboratory" for his ideas on economic measurement and social reform.[69]

Cheysson integrated social-scientific and hygienic worldviews, as an engineer and a social reformer. His concern with workers and wages derived from his eugenics. He feared the degeneration of the French population due to low birthrates, high mortality, and, above all, moral decline. Following Le Play, Cheysson identified the family as the center of the social fabric. He integrated Francis Galton's eugenic thought with Le Play's ideas on family and inheritance.[70]

Cheysson was one of many statisticians, including Lucien March of the Labor Office, with ties to pronatalist and eugenic groups. Statisticians created the data on birth and death rates, which motivated the founding of the French eugenic movement in the early twentieth century.[71] With statistician Jacques

Bertillon, who was also active in the French Eugenic Society, Cheysson co-founded the pronatalist National Alliance against Depopulation.[72] Cheysson viewed the working class as an organic social group—within a larger organic whole, the national population. To cure the population's ills, he argued, employers must care for their workers.

Cheysson warned that four great scourges threatened the French population: "the fever of needs, the difficulty of life, alcoholism and unhygienic housing."[73] In his many capacities he addressed each of those fronts. He directed a workers' housing development in the 1880s and organized a section on workers' housing at the 1889 Universal International Exhibition in Paris.[74] He joined hygienists in their campaign to moderate what he saw as the sources of workers' decline, their immoral and unhealthy behavior. In the realm of wages he sought to combat the pernicious desires that threatened workers' living conditions. Excessive needs were, literally, an illness. As one military doctor put it, "Only by fixing wages in proportion to real dietary needs, will we be able to combat the modern plagues of alcoholism and tuberculosis."[75]

For Cheysson, wage policy offered an instrument in combating population decline and degeneration. Cheysson found a way to navigate the wage question without offending liberal principles of the free market. His solution was for employers to supplement wages in order to support the lives of French families. He separated the wage into two parts: one set on the free market and another that supported the vital needs of the French family.

Cheysson warned against the danger of "false needs." Workers were "better clothed, housed and fed than before"; yet, he asked rhetorically, "are they any happier?" As their standard of living rises, Cheysson complained, workers "no longer derive contentment from the old sources of satisfaction." Modern workers required "dress shirts, handkerchiefs, and a thousand detailed items that were once reserved for the comfort and luxury of an elite few, but now are strict necessities even for the most humble existence." But even this, he warned, was not enough. Humble families longed for bourgeois pleasures. They "suffer . . . to see others drink long drafts from the cup of pleasure, and not be able to put it to their own lips." Cheysson worried about "the growing gap between reality and aspirations," which led workers to feel only "irritation, antagonism, and suffering."[76]

Cheysson feared that the draw of consumer pleasures and high nominal wages would lure peasants from the simple satisfactions of country life. Rural inhabitants did not have access to the temptations of monetary wages. They received much of their revenues in kind; "[rural] wages go straight into [satisfying workers'] needs, without being translated first into money."[77] Urban workers' wages, by contrast, "can be wasted in less pressing and less legitimate satisfactions."[78] Cheysson encouraged industrial employers to replicate the rural economy, to pay their workers with goods like bread instead of money. "[With] wages in kind [employers] can create a hierarchy of needs and satisfy the most essential ones, without allowing the worker to sacrifice them in favor of secondary or guilty needs." If employers paid workers with housing, heating, clothing, and food, no part will remain for the cabaret.[79]

Through the blunt instrument of wages in kind, Cheysson began to think about how to separate wages from needs. Wage policy derived not from economic calculations or state regulation of the "free" wage contract. Instead it flowed from a hygienic effort to maintain the quality and quantity of the national population. Cheysson integrated wages into a medical vision of population, life, and death. His amalgam of hygiene, pronatalism, and wage policy created a model for the modern French welfare state.

Cheysson's "fever of needs" clearly echoes a better-known social theory of consumption, articulated by the sociologist Emile Durkheim. Durkheim, like Cheysson, employed medical language to describe working-class consumption. Workers had been struck by an illness of modernity, the boundless impulse of unconstrained desire. In his 1897 *On Suicide*, Durkheim called it a "crisis of anomie."

In the modern world of industry and commerce, Durkheim warned, human desires had broken loose from any moral rule or reason.[80] Economic and technological progress did not guarantee human happiness. Rather, they risked unleashing uncontrolled and hazardous impulses.[81] Real happiness came not from passing pleasures, but from a balanced "relationship between needs and means."[82] This balance had been lost. Human needs were expanding without limit, leading inevitably to crisis and collapse. "Feverish imaginations . . . thirst for novelties, unknown pleasures, unnamed sensations, which lose their taste once they are known."[83] Durkheim decried such an insatiable search for pleasure as a "pathological" symptom of social failure.[84]

Individual happiness, Durkheim believed, depended on social cohesion and a collective hierarchy of needs. "How," he asked, "to set the quantity of well-being, of comfort, of luxury, that a human being can legitimately seek to obtain?"[85] Physical nature, man's organic body, could not provide an answer. Only among animals who act "automatically" by instinct, he suggested, can nature set the limits of need. Physiological measures could not explain why some professions and stations in life afforded greater needs than others. Society, not nature, must set a barometer for human needs by the "force of its moral power."[86]

At every stage in human history, men established a common moral standard of the revenues and comforts appropriate for each level of the social hierarchy. By custom and consensus, a specific lifestyle [*manière de vivre*] was assigned to each profession. "The worker in the city and the countryside, the household domestic and the day laborer, the commercial employee and the functionary" each held a place on the scale and risked opprobrium if he slipped too far above or below it.[87] A conventional scale of needs prevented dissatisfaction, especially among the poor, by constraining desires within limits that changed only imperceptibly. Social difference derived not from nature but from the disciplinary force of norms and habits.

Industrial progress and the proliferation of goods, Durkheim warned, had sundered the power of social norms. Economic growth "freed industrial relationships from any form of regulation."[88] Unbounded desires had become a regular and chronic cause of social maladies. They induced speculation and risk taking. They led to mental fatigue and economic ruin. If modern passions were not controlled by some external force, they became "a source of torment. . . . An unquenchable thirst is a constant torture."[89] Without social structures to contain it, the modern subject was condemned to a convulsive sickness of desire.

*

The late nineteenth century saw a flourishing of rhetoric on the subject of "false needs" (*besoins factices*). This notion had a long career as a specter of eighteenth-century thought.[90] Some so-called needs, conservatives asserted, simply should not count. Workers' consumption ought to keep pace with economic conditions, not with their own ever-expanding desires. What is "necessary," the conservatives argued, must correlate with what is "available": a lifestyle supported by current wage rates. They did not mean to allow wages to slip beyond employers' comfort margin. In the 1880s, economists and employers joined in a disciplinary campaign to control the apparently excessive expansion of workers' needs and desires. "Besides those material and immediate needs, which generally cost least to satisfy," one author claimed, "each man creates a crowd of needs for himself that are superfluous, or nearly so. Men definitely have more or less the needs that they want." Given free rein, "needs could grow beyond all measure."[91]

The economic crises of the late nineteenth century profoundly altered French society. For the first time in French history, the urban economy split from the rural economy. Urban markets were flooded with cheap imports of grains and meat. Crop prices plummeted, and French rural producers suffered, even as industrial workers gained access to a greater range of goods. Agricultural workers abandoned their fields and sought industrial jobs in the city; over three generations (1870s to 1910s), rural France lost one-fifth of its population.[92] Industry grew autonomous from agriculture; in this period both suffered parallel but separate crises. French industries no longer relied on rural consumers. Instead they floated and sank with urban demand and the international market for exported goods.[93]

French elites reacted to these developments with a concerted effort to manage consumer behavior. Industrialists and landowners allied to form a conservative republican bloc, which dominated a generation of French politics.[94] In the 1880s they succeeded in imposing heavy tariffs on imported agricultural and manufactured goods. This put an end to the free-trade experiment begun a generation earlier and impeded consumers' access to cheap imports. A significant portion of consumer spending was redirected toward domestic products.[95] In parallel, conservative social reformers pushed to impose moral limits on workers' consumption. Large industrial employers in this period became involved to an unprecedented degree in the consumer habits of their employees.

The rhetoric of false needs served many authors warning against overgenerous wage concessions. A 1885 work entitled *The Practical Economist* warned that "wage increases are often more harmful than useful to the interests of the working classes, since the worker does not know, in general, how to employ his money properly."[96] Even Fourierist sympathizers considered that "to increase the [worker's] wages would awaken a taste for expenses in him and would make it easier [for him] to satisfy it."[97] Improper predilections inevitably would siphon away the worker's nominal gains; therefore, for his own good, the wage ought to remain at a controlled level. This harsh assessment intruded into the otherwise celebratory context of the 1889 Universal International Exhibition. A secretary for the exhibition's "social economy" section concluded that the nation's century-long rise in wage rates stemmed above all from "needs that have increased, some legitimate and others false": "Nothing is more legitimate than a worker's family in search of well-being, healthier food, more appropriate housing, more dignified clothing; but beside this desirable progress, inspired by development and instruction, how many false needs have been created, aroused and maintained by contact with cities and big industrial areas, expensive needs which the wage is not intended to satisfy and for which the wage cannot suffice!"[98]

Henri Baudrillart, chair of political economy at the Collège de France, held a strict opinion of excess. He tempered an optimistic picture of ever-expanding civilization with a warning against "dishonest needs," defined as "needs that [in themselves] are not so false, but carried to an extreme." For Baudrillart, false needs resulted from "the competition between people of unequal condition over who can be most prodigal [*concurrence de prodigalité*] and the emulation of amour-propre, spread across all the classes."[99]

Baudrillart further suggested that this problem was a cultural and social, as well as individual, phenomenon. Needs, he wrote, are inculcated only "by habit."[100] His primary concern in that arena lay with the spending patterns of the lower classes. "How frequent is the pathetic abuse of luxury that appears when there is disproportion between expenses and income! How many people forget that even when an expense is not immoral nor strikingly excessive, by sacrificing the necessary to the superfluous it can become essentially blameworthy."[101]

Merry Delabost, professor at the Rouen School of Medicine and chief doctor for the French prison system, thought the same of dietary requirements. Why did Europeans eat more meat than other cultures, he asked? The only possible answer he found for such a divergence was bad habit. Delabost reproduced statistics on typical diets in English and Welsh prisons, of soldiers during the Franco-Prussian War, and of various professions and nationalities ranging from a London dressmaker to a sheepherder in the Urals. In a slight against his adversaries the hygienic "philanthropists," he denigrated the use of such data to quantify a universal maintenance ration. The variations between professions and regions, he contended, proved that civilized man overeats and stuffs himself with "false needs."[102] Urban workers should ignore the dictates of their desires and emulate vegetarians, peasants, and colonized races. "Can one live on ten sous a day?" asked Onésime Perrine, the author of *Healthy and Economic Living.* Yes indeed, "one can live for even less. . . . The price of food for myriad peoples in Africa and Asia does not rise above four centimes."[103] The implied equivalence of French workers and colonial peoples, a gesture common to nineteenth-century conservatives, elided class and race.[104]

Some dietary scientists sought to demonstrate that human bodies could adjust to ever more meager inputs. The key, they argued, lay in a gradual reduction of meal size. Subjects could "train themselves to live on little" as the body "spontaneously adapts to the amount of food which it is offered."[105] One army captain worried that the daily rations provided to every conscript would spoil rural recruits' simple tastes. "What happens to a peasant from the central plateau of the Pyrenees, accustomed to eating mostly lard, milk and potatoes, after two years of a meat-based diet? Quite often he abandons his fields forever and moves to the city."[106] Parisian workers in particular were accused of excessive consumption of meat and alcohol.[107] Louis Landouzy, dean of the School of Medicine and future member of the French consultative committee to the First International Eugenic Conference in 1912, castigated French "faults against rational nutrition" and encouraged schools to take up dietary education.[108] In the same stroke he argued for equal pay scales for women and men, on the basis of a reduced and equal set of needs.[109] Rational diets and wages would remove the false dictates of appetite and replace them with smaller, "real needs which regulate nutrition."[110]

Economist Edmond Villey also worried that workers were habituating to increasing levels of luxury. "Workers today consume necessary items in greater quantity than in the past, and enjoy consumption of luxuries that were once unheard-of; this fact is patently evident to us all."[111] Villey remarked that workers' standard of living could be "improved" in one of two ways: by raising pay rates (impossible) or by reducing their level of consumption. For workers to ameliorate their condition, "it is absolutely necessary that a change takes place in their ideas, their mores, and their manners."[112]

"False needs" played a central role in the debate among Catholic ecclesiastics and laymen following the publication of Pope Leo XIII's encyclical *Rerum Novarum: On the Condition of the Working Classes* in 1891. *Rerum Novarum* appeared as part of the pope's effort to realign the church with bourgeois republicanism; it also offered a defense against the threat to religion and private property represented by the working class's attraction to socialism. Leo XIII posited that a moral rule underlay and preceded the wage contract: "Let the working man and the employer make free agreements, and in particular let them agree freely as to the wages; nevertheless, there underlies a dictate of natural justice more imperious and ancient than any bargain between man and man, namely, that wages ought not to be insufficient to support a frugal and well-behaved wage-earner."[113]

In the wake of that papal pronouncement, even theologians entered into the business of measuring minimum standards of living. Catholic writers considered how to fulfill Leo XIII's requirement for commutative justice without imposing fundamental changes on the current wage system. Some argued, like the Abbé Naudet, that wages should precisely and entirely provide for needs: "mathematically, wages = life."[114] Social Catholic activist Henri Lorin expounded a Christian theory of labor and claimed that *Rerum Novarum* led inescapably to a "vital wage."[115] For this Lorin was bitterly denounced as "dangerous" by more conservative Catholics.[116]

Others like the bishop of Nancy hedged Leo XIII's apparent generosity, claiming that a "just wage" should only apply in proportion to the work that an individual performs. Any further payments for family support and basic needs, he argued, should count not as "justice" but as "charity."[117] A Father Jardin worried "how to set an exact figure that will provide the worker with enough to live. . . . Is it to live modestly, or rather in a comfort that leads to new requirements every day?"[118] Catholic journalist Charles Perin echoed that sentiment:

> It would be a serious error [*inconvénient*] if some exaggerated and inexact estimates [of needs] were allowed to remain in place, for they distort our understanding of the relation between the wage rate and workers' condition; these [estimates] would have us consider a given wage as absolutely insufficient even when it is only low due to the fact that people make bad use of it, to satisfy false and sometimes reprehensible needs.[119]

Father Jardin wondered rhetorically, "Where do real needs end? Where do false needs begin?" He answered his own question by example, with a set of goods that appeared beyond the threshold of need: "Is it necessary for the working family's budget to include a season at the seaside, evenings at the theater, cigars and absinthe?"[120] Clearly not. The range of these outrageous expenses suggests that Jardin measured the "falsehood" of needs with respect to the pleasure that they might procure.

Lurking behind these warnings against excess lay a hatred of public gathering places and alcohol.[121] Several authors of that period even went so far as to suggest that a policy of temperance for the working classes would solve the poverty problem all by itself. Deputy Auguste Bouge, a conservative Catholic from the Bouches du Rhône, asserted that coffee and wine were to blame for the flourishing of workers' "superfluous needs."[122] "If the worker did not drink," wrote Villey, "and saved all that he would have spent at the cabaret, we affirm without hesitation that most of our workers would live in relative comfort."[123] A worker seeking a bottle of wine, afternoon at the beach, public entertainment, or other pleasurable consumption perpetuated the dangerous view that needs could extend beyond bare subsistence.

VITAL WAGES

Employers at the turn of the century enacted a wide range of programs—including company stores, mutual insurance, education, and housing projects—designed to manage and direct workers' consumer habits. Women were constantly admonished to act as good "consumer citizens." Bourgeois women's magazines taught domestic accounting and how to serve both family and nation through shopping in "elegant economy."[124] Reformers castigated workingmen for drinking wine at the cabaret. Inspired by Cheysson, industrial employers implemented a bifurcated wage system that allowed them greater influence over workers' use of their money. The new wage regime distinguished between payments for work, which were set at the market rate, and payments for predetermined family needs.

Cheysson worried that most employers viewed the wage as a calculus of profit and loss, whereas workers saw it as their means of survival. To solve this imbalance he proposed that employers implement a broad range of charitable "subsidies, that is to say donations [*subventions, c'est-à-dire des libéralités*]."[125] These donations might emanate from employers, but also from communes or workers' associations. In Cheysson's mind they would cover a gamut of new and traditional goods: "coal allowances, schooling, medical services, retirement pensions, all kinds of aid; . . . indemnities for sickness, accidents, unemployment or old age; finally, in the countryside, grazing rights and the right to scavenge [*glanage*]."[126] In other words, welfare in its broadest sense. Cheysson's version of welfare, however, was not a right but a charitable donation. Cheysson campaigned heavily among large employers for a need-based family allowance. To employers interested in social peace and productivity, he urged a strategy of "going beyond the wage" to manage workers' needs.[127]

Textile factory employers in northern France were the first private firms to adopt the strategy of supplemental payments in the 1880s, followed by banks and service companies for water and gas.[128] Several state administrations, the Marines, Colonies, Education, Customs, and Finances among them, did so at roughly the same time.[129] To eliminate the potential for companies to use family allowances to undercut each other, a textile employer in Roubaix innovated the institution of an inter-employer fund. Managed by an employer-dominated mixed union or by employer representatives, industry-wide funds paid all workers their family allowances in equal measure.[130]

The textile business, under pressure from foreign imports, used family allowances to retain workers and to penalize strikers. Other industries, metals foremost, used the supplements to control workers' wage demands.[131] As Susan Pedersen notes, "Family policy is often wage policy in another guise."[132] Moralists downplayed the instrumental motivations for family allowances. Instead they celebrated their ability to "satisfy legitimate and incompressible needs, while avoiding the exaggerations which a uniform wage increase would not fail to bring about."[133]

A legal regime emerged for family allowances, which distinguished them juridically from wage payments. Although the allowances first appeared as additions to regular wages, employers and state officials soon took measures to distance the two. If family allowances had been given the legal status of an extra salary, they would have come under the jurisdiction of wage regulation. Allowances would then have counted in calculating severance, indemnity, and insurance payments. Family allowances would have been recognized as a legally defensible right. Instead, allowances were assigned the legal status of charitable donation [*libéralité*], revocable at will.[134] The modern welfare system grew out of these private initiatives by employers to refashion the balance between needs and wages.[135]

During the First World War the French government implemented a minimum wage in industries across the country. As the country mobilized for war, a vast realm of industrial production came under direct control of the state. Metals, munitions, and military supplies all entered the jurisdiction of the public purse. In its role as monopsony purchaser the national government closely oversaw and regulated working conditions in this wide swath of the economy. Within a short period, the provisions of a state-mandated minimum wage covered over two million French workers.[136] Economic instability required wartime workers to sacrifice their well-being, while at the same time the war "produced a powerful logic to redefine the notion of citizenship in terms of a minimum wage."[137]

Massive strikes protested the high cost of living.[138] In the north these actions were amplified by food riots, demonstrations directly concerned with the price of bread and other staples.[139] The Labor Ministry estimated that the cost of a basic family budget rose by 40 percent from 1914 to 1917; by contrast, average wages rose by 20 percent in the period from 1911 to 1916. The result was a sharp descent of real wages in most French professions and regions.[140] Pamphlets and newspapers copied sample budgets of postmen, teachers, office workers, and heavy laborers. "The best way to give an account of the increasing cost of living, and to compare prices and wages, is without a doubt to draw up current budgets for workers' or functionaries' families."[141]

Miners and steel workers' unions were at the forefront of the movement for cost-of-living adjustments, which reached its peak in the mid-1920s.[142] Henri Prêté, secretary of the Committee on Action of Workers' Organizations, echoed a common refrain in his call for a legal minimum wage. "The basic reasoning behind the current [workers'] movement and its demands does not depend on the whim of the unions; it stems from the imperious necessity to

FIGURE 9. Billancourt, France: August 1, 1917. World War I. Albert Thomas (1878–1932), minister of ammunitions, making a speech at the Renault car manufacture in front of 23,000 workers. Photograph published in the newspaper *Excelsior* on Sunday, September 2, 1917. © Excelsior L'Equipe / Roger-Viollet / The Image Works.

obtain what is indispensable to live." Prêté grounded workers' wage claims in the inexorable nature of need. "It is a pressing necessity to raise wages, to re-establish equilibrium between daily earnings and the value of things necessary for life."[143]

Workers' newspapers and union pamphlets repeatedly proclaimed their human right to a decent living. Casimir Bartuel, secretary of the Federation of Miners, produced an exemplary report for the 1918 Congress of the Federation of Underground Workers.

> In our society, whose existence the worker ensures and whose wealth the worker produces by his labor, does the worker have a right to remuneration in the form of cash value? Does he have a right to exist? Who would dare to pretend that he does not have that right? It is a natural, humane, logical, imprescriptible right. . . . What should be the point of departure for the application of this principle? The quantitative evaluation of representative values corresponding to what is necessary for the daily life of the worker and his family.[144]

Bartuel undertook a statistical study in the Isère region to determine "the price necessary for a working household to live modestly, taking into account the intensive physical effort asked of miners." He composed a sample weekly budget based on a simple diet of bread, meat, potatoes, dried vegetables, cheese, eggs, sugar, oil, coffee, and wine, and he calculated the relative cost for the same items in 1914 and 1917.[145] To this basic diet Bartuel added the cost of rent, clothing, heating, furniture, and savings. He found a nearly threefold increase in the price of this package over the course of the war. "The quantities indicated in this table," he wrote, "constitute a strictly indispensable [amount], and their market value corresponds exactly to the minimum wage to which every worker in the region where these prices were collected has a right."[146] Bartuel demanded that the state enforce this strict proportionality.

The Labor Ministry established commissions in war industries to set need-based wage standards. State-controlled monopolies in tobacco, matches, and gunpowder, and the national mint, all guaranteed a minimum wage; pay rates were proportioned to local cost-of-living statistics.[147] Local councils of *prud'hommes* sent the ministry statistics on average wages in each town and region. The Labor Office compiled a massive collection of figures on average pay rates, workers' budgets, and the cost of basic goods.[148]

When the state began to measure and implement a standard regional minimum wage in its own establishments, pressure increased for private employers to follow. Minimum wage clauses became standard in contracts between trade unions and employers. As a result of these developments, wages across all industries became more uniform. The lowest-paying employers conformed to a more average rate, and the range of pay scales across the country grew flatter.[149] The minimum wage gradually became a unitary national standard, based on a "vital minimum."[150]

Following a particularly intense period of strikes during the winter of 1916–17, Labor Ministry officials adopted a new wage-setting strategy: the "vital wage." They divided the wage into four components: a base wage (*salaire d'affûtage*) would be awarded to all workers regardless of seniority or skill. A second sum could be added on the basis of merit or speed, using a piecework rate. These first two rates were to remain stable, regardless of the cost of living. They were standardized by industry and tied directly to production.

Two further elements would compose the need-based part of a worker's income. The cost-of-living supplement (*prime de cherté de vie*) allowed for a continual adjustment of pay to account for inflation. This supplement remained the same across all industries and varied according to region and gender. Adjustments to cost-of-living pay would take place automatically based on a standard budgetary consumer price index. Finally a family allowance (*allocation familiale*) applied only to male workers and was adjusted to the number of his progeny.[151]

The resulting separation of needs and productivity allowed employers to maintain the wage constant while placating workers' demands with various "supplements." The base rate, designed to apply to "non-specialized manual laborers," was intended to subsidize no more than pure survival. The supplemental components of the wage, on the other hand, were now mechanically linked to a standard consumer index. A statistical measure sheltered family supplements from negotiation and challenge. The need-based family allowance was thereby officially separated from the wage.

Labor Ministry officials designed the vital wage to palliate multiple constituencies. "Vital" evokes a medical, hygienic measure and also the Socialist "iron law." Officials drew upon pronatalist language to justify family allowances, and they also indexed allowances to consumer budgets in an echo of previous wage legislation. The merit-based components allowed a margin for employers to impose a productivist agenda, tying pay to performance. Other components referred only to needs.

Was the vital wage a physiological or a social measure? Officials defined the "vital" quality of a wage by both measures, even to the point of contradiction. William Oualid directed wage policy at the Labor Ministry during the war and implemented the vital wage. He wrote extensively about wage policy during and after the war. In a striking reference to nutrition science, he described the base wage as "the *minimum price of a maintenance ration* for a worker with average individual and family needs."[152] One of Oualid's students similarly defined "vital wage" as "a minimum, which strictly ensures the conservation of life." The vital wage, in this model, appears as an analogue to the "necessary wage" that Jules Guesde had criticized. "One must admit that beneath a certain minimum level of food, the human body is exhausted and life is shortened"; this limit defined the vital wage.[153] Some workers at the time perceived the vital wage as an absolute minimum. An article in the *Cahiers du socialiste* described the vital wage as "the lowest possible sum, beyond which an employee can no longer ensure his existence."[154]

Elsewhere Oualid extolled the vital wage as a "social" measure, meant to promote a "policy of workers' well-being." The vital wage, claimed Oualid, guaranteed that the "average worker" could provide for himself and his family, and "occupy the social rank that he is due."[155] "The social, or vital, character of the wage" was "based on the needs of a worker and his family and not only on his technical productivity." Oualid cited family allowances as the best example of this "social, vital" view of wages.[156] The "vital" in "vital wage" thus referred to both the body and the family. It guaranteed the reproduction of labor force and the reproduction of the laborer's family. It melded a chemical-physiological view of the individual body with a social, collective view of the national body. In this sense it was a logical product of its time.

The most colorful critic of the vital wage and family allowance system was undoubtedly Hyacinthe Dubreuil, a locksmith, *compagnon,* and leader of the Confédération Générale du Travail union. Dubreuil is best remembered for his conversion to Taylorism during a period of employment abroad at the Ford factory. His *Standards: American Labor as Seen by a French Worker* (1929) had great success in France and even was published in the United States under the title *Robots or Men?*[157]

In the early 1920s, however, Dubreuil occupied himself primarily with a scathing critique of the need-based wage. "Over the past few years," he wrote in 1923, "especially since the war brought on considerable economic unbalance, Cost of Living commissions tranquilly have begun to establish that a worker needs so much bread, fat or shoes, or so many movies, or even so many calories, to survive." Dubreuil denounced these measures as a "mechanical" form of "barbarity" which renders workers' "sweat and blood, their entire lives, in algebraic expressions."[158]

> Ever since this magnificent method was introduced, some absolutely stupefying arguments can be heard in wage negotiations between employers and workers. Just recently one of the more notorious of those employers, from an important company, squabbled like [Molière's curmudgeonly character] Harpagon about the amount of meat that a worker should aspire to! . . . To see such a serious discussion fall so low is sickening.[159]

Miserly employers had managed to outfox and mystify their opponents. "A great number of worker activists have allowed themselves to be hypnotized by this odious arithmetic, and naively agree to argue about the elements of its equation." Workers' unions do their adherents a grave disservice, Dubreuil argued, by adopting quantitative measures of needs and the cost of living. Such calculations mask the social worth and dignity of work, which the author considered the real measure of its value. In anticipation of his later writings on Taylorism, Dubreuil called for worker participation in factory management rather than a need-based wage negotiation. "We should raise the level of debate . . . to more honorable arguments than the proportion of beans that are swimming in our meager soup!"[160]

Dubreuil spent the following two decades in search of an alternative to the profit-oriented republican economy that spawned such a humiliating logic. In the 1930s he joined with like-minded engineers and social scientists in quest of what engineer Jean Coutrot called "economic humanism." Their economic vision displaced profit and speculation in favor of authentic human needs. Coutrot's Center for the Study of Human Problems, established in 1936, attracted biologist Alexis Carrel and the English writer Aldous Huxley, psychoanalysts, work scientists, and labor movement leaders including Dubreuil and future Vichy secretary of labor René Belin.[161] Like many others, Dubreuil saw the French defeat in 1940 and the "National Revolution" as an opportunity to realize profound economic and social reform. He dedicated a book on the *Chivalry of Work* to Maréchal Philippe Pétain.[162] Hope in the new regime faded rapidly, however, and Dubreuil retreated to Paris in January 1942.[163] Dubreuil's critique of the republican economy forms part of the next chapter's story. The human, social qualities of consumption disappear, Dubreuil argued, in simplistic equations of budgets and pay rates. "One might as well say that we scientifically determine a worker's rations like we determine rations for farm animals!"[164]

CHAPTER 7

Science of Man

In states across Europe, economist François Perroux rejoiced in 1944, "biology is erupting into politics."[1] Everywhere, Perroux declared, "the profit economy is receding, to the benefit of an economy that satisfies needs, objectively evaluated [*apprécié*]."[2] Biologists and human scientists offered the foundation for a new social organization. "If [science] knows the order of human nature, it can indicate the minimum to which producers, consumers, and their leaders must submit."[3] The scientist's work, far more than a collection of statistics, was to reconstruct an authentic, "real" human. Science would lead the way toward "an economy ordered according to the ends of man, and evolving in the direction of human life."[4] Scientists were helping to reset the course of nations.

The Vichy regime of "National Revolution," formed in the shadow of German occupation, embraced Perroux's vision. "Give the people their vital minimum," President Pierre Laval ordered French prefects on the title page of *Aujourd'hui* in January 1943.[5] "Quick, the minimum wage!" called a state-sponsored workers' newspaper in May 1941. "Here are the strictly minimum expenses that are indispensable for all of us, the vital minimum."[6] A Vichy government press release in 1941 vaunted "an innovation that one can qualify as truly revolutionary: the establishment of a vital minimum wage [*salaire minimum vital*]."[7] Vichy enshrined a scientific measure of basic needs, the measure of a vital minimum, in the human sciences and in national law.

For many French social scientists, the disaster of defeat in 1940 was exactly what the country needed to jumpstart an economic National Revolution. Constrained by a precipitous drop in production and steep German levies, Vichy officials had no choice but to assert tight state control over the market. Though some regions experienced far more painful scarcity than others, food shortages impacted almost all French residents. In response, authorities strictly regulated the market for consumer goods. "Instead of a distribution of goods based on purchasing power," wrote an economist close to Vichy, "we want to substitute a distribution predicated on needs."[8]

Here I am interested in the scientific vision behind Vichy's authoritarian welfare state, a hybrid of economic theory and sociobiology. A turn to the archive reveals how science and politics shaped two of Vichy's major welfare policies: food rationing and the minimum wage. These stories provide material for historians to debunk the exceptional nature of the Vichy regime. Vichy social thought drew much from the Third Republic and formed much of what would become postwar France.[9] More broadly, this story contributes to an ideological and scientific history of economics and welfare. The tools of neoclassical economics, such as general equilibrium and marginal utility, proved flexible enough to adapt to dirigiste, authoritarian states. The twentieth-century European welfare state was as much a racial-eugenic regime as a social-democratic one. Welfare logics and technologies passed from right to left and reached a historical apotheosis in the authoritarian regimes of the mid twentieth century.[10]

Wartime France was an "immense human laboratory in which medical and psychological experts manufactured massive amounts of sociobiological data."[11] The Vichy government employed more policy-oriented scientific researchers than any previous French regime.[12] State-sponsored scientists investigated French citizens' food, budgets, housing, productivity, reproduction, and health. The aim of this accumulated data was to create a definitive profile of the population's quality and needs. To this end Vichy financed social-scientific research at record rates. Studies once carried out by para-political groups, such as the Musée Social or pronatalist, pro-family associations, now emanated directly from state institutions. The science of man stood at the center of state policy.

The Vichy science of man was empirical, based on extensive observation and data from a variety of sources and disciplines. This science emphasized social differences, not universal abstractions. Research would be "concerned only with the study of the 'real' person, that is, the 'concrete' human being in its specific setting—'the worker in the factory, the child in school, the housewife in the kitchen, the peasant on the farm, the alcoholic in the cafe.'"[13]

Vichy social science developed along three lines, each drawing from a long French history. The first, the corporatist orientation, collected market data on prices and wages. This project drew directly from the trade union statistical tradition. Although endorsed by Vichy labor officials and newly created professional corporations of employers and workers, this approach had only a minor impact on policy. A second current drew from nutrition science, employing chemical and statistical measures. This, too, received relatively little attention from the state. Last, and far more powerful, a sociobiological orientation followed the lines traced by Gérando, Le Play, and Cheysson. This research program sought human needs outside of the market, in studies of biological and racial types. Like nineteenth-century conservatives, Vichy scientists built racial-biological hierarchies of needs and capacities. This time, the state put these hierarchies into practice on a national scale.

BIOSOCIAL ECONOMICS

François Perroux's life and work were marked by economic depression and war. He was deeply suspicious of market values, for which he wished to substitute a national, communitarian economy. State authorities would direct the economy according to human needs, "objectively measured." This vision came close to realization in the early years of the Vichy regime. We can view Perroux as an organic intellectual of Vichy economic policy.

Perroux is interesting to the history of European economics as a kind of mirror. In his early economic work he interpolated theories of marginal utility and American institutionalism. By 1935 he arrived at a synthesis sympathetic to Italian Fascist corporatism. Following the French defeat and German occupation, Perroux held an influential role in Vichy economic planning. In 1941 he was named to the Constitutional Commission of the National Council and to the Council for Economic Study.[14] From 1941 through 1943 he served as secretary general of the state-financed, eugenicist French Foundation for the Study of Human Problems.[15] After 1945 he would perform an analogous synthesis of Keynesianism and spatial theories of development. Perroux, as Julian Jackson remarks, fitted neither a "modern" nor a "reactionary" mold, but moved fluidly between concepts that fitted the age.[16]

Economists in the early twentieth century understood their discipline as the study of distribution under conditions of scarcity. They sought to understand what rules governed distribution of scarce goods. This orientation toward the consumer represented a clean break with earlier political economy. Unlike Adam Smith or Marx, who put forth a theory of value based on labor, modern economists focused on consumption and distribution. The English economist Lionel Robbins, in a paradigmatic phrase, summed up economic exchange as "the forms assumed by human behavior in disposing of scarce means."[17]

Neoclassical economists, inspired by the work of William Stanley Jevons, Carl Menger, and Léon Walras in the 1870s, sought to model how markets functioned to distribute scarce goods. Jevons, Walras and Menger were concerned with the satisfaction that a consumer gets from each new unit of a particular good. How much satisfaction would a person get from five loaves of bread, as opposed to four? Or six, as opposed to five? The answer to such questions could be used to determine how markets function. When the satisfaction, or utility, of consumers in a given market began to decrease, the price commanded by that product would fall. In an efficient market, just enough bread would be distributed so that the marginal utility of all bread consumed would equal the marginal cost to produce it. The cost to produce the very last loaf of bread should be equal to a consumer's utility in eating it up. Economists have since labeled this the problem of marginal utility.

Jevons inherited the term "utility" from a British lineage going back to Jeremy Bentham. Bentham proposed a measure of units of sensation, pleasure, or pain. If one could add up a sum of pleasures and pains, one would reach an overall measure of the utility of a particular good or policy. Jevons used utility in a limited sense, to define how markets distribute goods. In the late nineteenth century psycho-physiologist Ludwig Fechner sought to give this unit an empirical basis. He attempted to measure the smallest possible change in human sensation. British economists following Jevons speculated that Fechner's work might provide a scientific measure of utility. But Jevons, like Menger and Walras, did not believe that this unit could or should be measured by economists.[18]

Utility meant more than a simple unit measure of pleasure or pain. Walras and Menger avoided Bentham's and Jevons's penchant for individual psychology. However they too were concerned with fitting goods to needs or wants at the margin of satisfaction. Carl Menger articulated a concept closest to the vital minimum, the "satisfaction of need." For Menger, a thing became an economic good only when it was capable of satisfying a legitimate and existing human need.[19] However Menger did not attempt to quantify needs, which "derive from our drives and the drives that are embedded in our nature."[20]

Walras was concerned with what he called the *rareté* of a particular good. He composed a complex mathematical model to describe how markets distribute goods. He proposed that an ideal, pure, and unencumbered free market would eventually reach an optimal point of general equilibrium, matching goods to needs. Both Walras and Menger argued that *rarété* or need satisfaction could be measured only indirectly, by recording the price that a consumer is willing to pay for an extra piece of a good. Walras compared this to physical measures of mass and force, which are indirect measures of an object's molecules and energy.[21]

Marginal utility theorists faced the basic problem that one person's utility or satisfaction could not be compared with another's. The price a consumer paid for bread could inform economists what the bread's utility might be to that consumer. But how to measure whether a loaf of bread would give greater satisfaction to one person than to another? No common, interpersonal measure existed. Italian economist Vilfredo Pareto took up this problem in the 1920s and overturned the notion that measures of utility could ever be comparable.

Pareto designed a purely subjective, personal measure. He observed that consumers rank goods on their own scales of need or desire. Economists need not worry about what those needs or desires might be, how they compared to others' needs, or whether they were socially legitimate. All one needed was to know whether a consumer preferred bread or pasta, pasta or rice, rice or polenta, and in which order. With Pareto, concerns about measuring needs and satisfactions gave way to the study of independent individual preferences, or ordinal utility.[22] As economist Gary Becker put it years later, economists should study consumers' choices, not their tastes and needs: "De Gustibus non est Disputandem."[23]

By the early twentieth century, mainstream economists had ceased to search for a standard measure of utility or satisfaction of need. Swedish economist Knut Wicksell expressed a sentiment common to his colleagues at the turn of the century when he rejected interpersonal measures of utility. Nonetheless Wicksell speculated that "perhaps someday the physiologists will succeed in isolating and evaluating the various human needs for bodily warmth, nourishment, variety, recreation, stimulation, ornament, harmony, etc., and thereby lay a really rational foundation for the theory of consumption.[24]

By 1941 François Perroux believed that such a measure had become possible. Perroux joined a group of European economists who attempted to reconcile neoclassical economics and a social, corporatist economy. Perroux, like many fascist and corporatist thinkers, sought to replace abstract mathematical models by "real, concrete" ideas. He drew from institutional economics to argue that economic theory must take account of a country's "legal, technical, social framework," "the national structure of an economic system."[25] Perroux's wartime economics rejected individual choice and subjective judgments of need. Instead he proposed a state-directed economy organized according to collective, national priorities. In place of individual preferences Perroux hoped to build an objective hierarchy of need. He had help from a long French tradition of human science.

Perroux was imbued with a profound distrust of the free market. His father, an artisan shoemaker, had lost his livelihood during Perroux's childhood due to tight competition.[26] He was left with a deep skepticism of market and monetary values. At several points during a remarkably long and prolific career, he admonished his readers not to substitute "money, the expression of goods" for "the goods themselves."[27] Perroux's quest for more enduring values also drew from his fervent Catholicism. He wrote frequently for Dominican journals and publishers, and he formed part of the "fraternal network" of antirepublican, anti-free market Catholics who would play an important role in the first years of the Vichy regime.[28]

Perroux spent the interwar years at the Faculty of Law in Lyon, a center of French social Catholicism. His mentor there was Etienne Antonelli, a former student of Léon Walras. Through Antonelli, Perroux made contact with the major European economists of the day. A Rockefeller Foundation scholarship to Vienna in 1934 consolidated intellectual friendships with Oskar Morgenstern, Hugh Gaitskill, Werner Sombart, and, most importantly, Joseph Schumpeter, to whom Perroux dedicated his first major publication.[29]

Antonelli taught Perroux to integrate economics into social, historical, and institutional context. Antonelli won election as a deputy in the 1920s in order to further new legal institutions—social insurance, professional confederations—that promised to better serve human ends.[30] His basic premise, and Perroux's, was that all human interaction, at its core, aimed to satisfy the most essential, vital needs. Perroux and Antonelli joined men from a range of ideological standpoints in the 1930s and 1940s who aspired to a "scientific humanism" grounded in the science of need.

The Great Depression exposed a capitalism seemingly out of whack, propelled by profit and speculation to the detriment of the common good. As many as 1.3 million French industrial laborers lost their jobs between 1931 and 1936.[31] Due to a combination of unemployment and expanding managerial ranks, wage earning began to lose its status as a marker of social class.[32] In 1936, massive strikes shut down the country and ushered the Popular Front to power. Faced with the brutality of the Great Depression, many became disillusioned with free market capitalism.

Young engineers and doctors, social Catholics, neosyndicalists, and fascist sympathizers, radicals on the right and left formed a loose group of "nonconformists."[33] These men blamed uncontrolled speculation and profiteering for the country's crisis. They were as repulsed by parliamentary politics and capitalism as by the Communist alternative. They abhorred Americans' unbridled mass consumption as much as Soviets' egalitarian rhetoric, and sought to chart a third way between them. Many of them drew their political ideas from the same sources as early Italian fascists: Charles Péguy and Georges Sorel. They shared a common distrust of monetary values, distaste for the profit motive, and desire for scientific solutions to economic ills. They sought to replace the liberal free market with a "human economy," a new form of social organization geared toward "real" people and their needs.[34] As two young Catholic leaders wrote, "The economy should be in the service of man."[35]

Perroux was deeply involved in the formation of a conservative Catholic network under Vichy. He gave several lectures at Uriage, a semi-public training school for future Catholic leaders, and he was a founding member of the Dominican reformist group Economy and Humanism.[36] Like many social Catholics, Perroux expressed increasingly vocal critiques of Vichy after 1942. He nevertheless remained a strong supporter of the regime until its end.[37] Perroux kept his expert position in the Vichy administration long after other antiliberal figures, like Minister of Economy René Belin, departed from government.

Perroux also became one of the most influential promoters of French corporatism. The corporation drew upon nostalgia for an imagined organic, harmonious social order centuries before the rise of modern capitalism and the French Revolution. Workers and employers would together build self-organizing social groupings based on collective and national identities, not social class.[38] Corporatism offered a third way between the free market and Marxism, a "socialisation without socialism."[39]

On several occasions in the 1930s and 1940s, Perroux held forth Italian corporatist Filippo Carli as a model for his new economics. Carli argued that individual needs and wants could not form the basis of a healthy and dynamic economy. He discarded the neoclassical *homo economicus* in favor of a new *homo corporativus*: a man who subordinated his individual interests and needs to those of the national body. Subjective feelings of desire or want should give way to objectively defined collective needs. Carli, who was for a time closely identified with Fascist policy, called his brand of corporatist economics a "pure national economy."[40]

Perroux's most important achievement during the Vichy period was to bring a racially inflected biological science of man into economic theory. From 1941 to 1943 Perroux served as the first secretary general of the newly created French Foundation for the Study of Human Problems and as head of its biosociology department. The foundation, described in more detail below, combined modern eugenics, racial science, and a conservative sociology drawn from Joseph de Gérando, Frédéric Le Play, and Emile Cheysson. In biosociology Perroux saw the answer to the problem of an "objective hierarchy of needs."[41]

In a series of manifestos published between 1942 and 1945, Perroux articulated the theoretical and practical bases for an "economy of needs."[42] Perroux's was an economy "ordered and measured by the needs of a living, concrete man, his body and soul."[43] It was an economy of rationing, basic welfare and the science of man: "When specialists and the larger public in the Nordic countries . . . stress that the division of income has no value in itself, but must be reinforced by social hygiene in its broadest sense, the economy is corrected by biology and the science of man."[44]

"Social hygiene" here resonated with the history of public health, sanitation, and welfare, as well as with German racial hygiene.[45] Perroux's economics did not place much explicit emphasis on the racial content of Nordic social hygiene, though he drew parallel conclusions. And the foundation certainly did have a eugenic mandate. In an economic brochure written for the foundation Perroux praised "public powers" for "taking care of the baby, the mother, the young, deciding to protect the biological patrimony, the blood and race . . . using mass methods. . . . Not since the ancient Cities has the cult of man's sacred flesh been so strong."[46] Perroux's "living, concrete man" was a man defined by blood and flesh, nationhood, race, and bodily health.

Social hygiene offered an alternative definition of use value, a "use without appropriation." Perroux listed fundamentals of nineteenth-century public health campaigns: fresh air, a park in the public square, sunlight in large windows. He equally applied the category of "use without appropriation" to "the happy conversation and *joie de vivre* of our fellow workers." Together these goods "constitute a kind of biological property—communal, not privately held—where "being" joins with "having."[47]

Perroux criticized abstract mathematical models of economic activity, which he claimed had erased the "living, total man from economic science."[48] Modern man, drawn by the sirens of marketing and "false needs," had become "foreign to his own nature."[49] In a profit-oriented market, a person's needs depended not on his nature but on his ability to pay. Perroux chastised Menger and Walras for having taken those conditions at face value, confusing people's means and subjective desires for their "natural needs."[50] Economic statistics, like those of Emile Durkheim and his student François Simiand, were a positive step toward a more "concrete man." However Perroux had little faith in the potential of statistical data, which consisted "only of averages," to reveal the true nature of man.[51]

Perroux aspired to a "synthetic economics," which combined the theory of marginal utility with a "concrete" analysis of institutions, political and economic systems.[52] This new economics would draw from two currents of scientific thought. First, Perroux pointed to Werner Sombart and German historical sociologists, who "go beyond the sociological point of view to build a general anthropology." The questions that Sombart and colleagues sought to answer would give clues to the order of human need: "What is man? What are the laws of man?"[53]

Second, Perroux summoned Alexis Carrel's eugenic work into the field of economics. Founder of the French Foundation for the Study of Human Problems, Carrel was a surgeon, Nobel laureate, and prominent eugenicist. Perroux was persuaded that the foundation would discover that "there is a natural order to the human species, that science can discover."[54] The natural order had been buried under centuries of social transformation. Only a deep exploration of biology and society would unearth it. Biosociology offered the key to the objective order of human needs, upon which a new order could be erected. "If [the human sciences] know the order of human nature, they can specify a [vital] minimum. Producers, consumers, and their leaders must submit to it. [The minimum] appears as a sure and certain guide to an economic science that is formal and neutral, and an agnostic economic practice."[55]

This economy, the economy of the twentieth century, would be an economy of rationing and the minimum wage. It would be the "communitarian and authoritarian" economy of war and the post-democratic state.[56]

*

François Perroux believed that the French Foundation for the Study of Human Problems would create a true science of man, the key to an objective hierarchy of needs. The organization's mandate reflected the gigantic eugenic ambitions of its director, Alexis Carrel.[57] Carrel dedicated his foundation to the reconstruction of man. He had expertise in scientific organization from his lengthy tenure at the Rockefeller Institute for Medical Research, where he had developed new methods of organ preservation and transplant. His political sympathies favored fascist elements of the Vichy government, and he had promoted eugenic thought in his 1935 best-seller, *Man, the Unknown.*

The foundation's organization indicates its scope. Its generous funding reflected the Maréchal Pétain's interest in its work: it received nearly as much—45 million francs—as Vichy's entire budget for laboratory research.[58] Research divisions included Biology of Lineage, Natality, Child and Youth

Development, Housing, Nutrition, Biotypology, Psychophysiology, Work, Rural Economy, Production, Sociology, Economy, Finance, Law and Insurance. In addition a Survey and Statistics group applied Gallup methods to survey public opinion. The foundation created voluminous data on population behavior, health, and attitudes, which it aimed to synthesize into a program for state action. Research sections covered "the French population's most urgent needs, both material and mental," thus nearly every aspect of family life and personal behavior, population quality and national productivity.[59] Every piece of this research plan responded to "national interests," the imperatives of state planning and policy.[60] Carrel compared the foundation's work with that of airplane engineers: he pledged to devise a science of "anthropotechnics," or "the construction of civilized men."[61]

The foundation combined a modern eugenic vision of biological difference with conservative French sociological tradition. In a project strikingly reminiscent of Joseph de Gérando's work 130 years earlier, the foundation sent investigators to small, isolated villages. They selected " 'homogenous' populations that were perceived as having an 'ancestral way of life' with hardly any exposure to 'external influences.' " Every detail of the inhabitants' lives was catalogued and examined, thought to reveal a human community free of capitalist modernity and its corrupting pressures.

Foundation researchers created 2,500 files detailing desirable and undesirable hereditary traits. This data served to create a tableau of hundreds of "human types."[62] This work fitted exactly within a eugenics science practiced across Europe and by the American Eugenics Research Office. Carrel and Perroux's use of this data also refers back to Emile Cheysson's ethnographic collection of racial types, distinguished by unique traits. Like Cheysson, foundation scientists saw themselves as social engineers.[63]

The "eruption" of biology and the science of man accompanied a surge in eugenic thought.[64] The French Foundation for the Study of Human Problems was the first French state institution to publicly endorse eugenics.[65] Scientific organisms under Vichy were preoccupied with maintaining population health and quality. The science of man defined in essence who was human and who was not. To measure "authentic human needs" was a fraught proposition in a time when Jews and Romanies were dispossessed of their homes, livelihoods, and even basic sustenance.

A large part of the foundation's work dealt with how to identify and satisfy authentic human needs. Foundation scientists rejected the utility of universal physiological standards. Instead, they sought to establish biological types, each with its own requirements. Nutrition belonged to the demographic section, alongside "birth rate, pathological sterility, heredity, homogenous groups . . . alcoholism."[66] The foundation began a "Study of Dietary Nutrition" in spring 1941. The study's authors disputed the claim that diet was related to social class. "If one wishes to know the diet of a group of individuals," they argued, "one must deal only with the food products that are absorbed. Physiological, economic, and financial concerns logically should be secondary investigations that later can be combined, or not, with the study's [primary] objective."[67] The facts of biology, for them, fit into a hierarchical classification, not a universalist physiology. Biological types, not class, should ground the distribution of goods.

In the course of its study, foundation interviewers asked passersby on the streets of Paris to reconstruct their menu for the last twenty-four hours. The survey form comprised only two pages divided into "breakfast," "lunch," and "dinner." A brief section for "Observations" recorded the participant's sex, age, profession, marital status, and appearance of health. Numbered codes permitted interviewers to note "thinness," "weakness and deficiency," disease, nervousness, or "general bad condition."[68] The researchers, despite their strict admonition to exclude social considerations, still included questions concerning the respondent's possession of a telephone or automobile, his or her arrondissement, and number of household domestic servants.

The foundation surveys revealed that salespeople, clerks, and domestic workers ate little. In contrast to industrial workers, low-level employees barely ever bought meat. On one day in mid-December 1944, for example, a middle-aged dressmaker in the Fifteenth Arrondissement ate only a few green beans, a bit of bread, and an apple for both lunch and dinner. Post-office workers and policemen, on the other hand, consumed relatively plentiful meals.[69] Despite evident gaps between social categories, the researchers preferred to attribute differences in diet to biological factors. Age and sex, not class or social hierarchy, determined French people's dietary regimen.[70] The foundation used its results to advocate in favor of increased milk for children and more rationally designed meals in school canteens and soup kitchens.[71]

RATIONING

The vital minimum, under the extreme conditions of wartime scarcity, meant food. François Perroux pointed to wartime rationing as an example of a scientific, rational economy coordinated by the state. Wartime penury left the French, despite their historically "anarchistic" temperament, no choice but to impose "general and objective rules" on consumer choices. Perroux praised the scientific character of the commission charged with establishing rations for children, adults, workers, and prisoners. Medical and economic experts worked together, he claimed, to set an objective scale. As a result war rations took into account the full range of "human conditions and repercussions," social as well as physiological measures. Perroux emphasized that Vichy did not simply extrapolate its rationing system from abstract caloric standards, but considered the full science of man.[72] A close examination of Vichy nutrition science and policies bears out Perroux's characterization of rationing as a project of social hygiene.

During the entire decade of the 1940s, both republican and authoritarian French governments imposed a strict and hierarchical rationing system. Wartime governments staked out expansive powers of economic regulation. The Third Republic, in one of its final gestures, required every resident to register at the local mayor's office for a rationing card by April 1940.[73] Minister Paul Reynaud justified upcoming consumer restrictions in a radio address using language inherited from the Siege of Paris of 1871: "Since scarcity is inevitable, rationing is the only equitable form of distribution."[74]

Rationing reached its fullest extent following the defeat and mass population exodus of May–June 1940, under duress from the German military administration. The Vichy administration expanded and reinforced the rationing system, which now included tickets and price controls for foods and over a dozen basic goods from soap to tobacco. "Bread, wine, sugar, milk, meat, oil and other fats, all preserves (vegetable, meat, or fish), jams, sweets, chocolate, coffee, potatoes, fruits and even vegetables . . . were all restricted at some point."[75] The Maréchal Pétain, like Reynaud, promised to provide both "rich and poor" with "sufficient nutrition," "a just part of the nation's resources," and "equality for all in sacrifice."[76] A local newspaper in Troyes picked up Pétain's message: "One must not criticize rationing overmuch; however imperfectly organized, it still insures that poor people receive a minimum that no one had previously dreamed of guaranteeing."[77]

Expansive state intervention and a tightly restricted flow of products dramatically affected the entire French population. Control over distribution and consumption emanated from a centralized Ministry of Provisioning. Police dealt harshly, though in vain, with counterfeiters and black marketeers. The government's policy, by necessity and design, was to prioritize between population groups, to measure needs and supplies. The state guaranteed necessary goods to those judged worthy and limited access to others. Scientific organizations participated in this dynamic. Researchers set standards for population quality and measured "real" needs.

During the occupation the state tightened its definition of the national community to exclude Communists, Romanies, and Jews. Government propaganda lining the back of "T" (worker) ration tickets showed hoarders and speculators with protuberant noses. "The Jew doesn't earn his bread by the sweat of his brow," the caption read. "He steals by speculating and trafficking the products of other people's work."[78] Soon after the German military authority imposed a yellow star on every Jew in France (in exchange for a textile ration ticket), Prime Minister Laval ordered in December 1942 that "JEW" be stamped on all applicable rationing cards.[79] Deprived by confiscation and exclusion of their means of livelihood, those Jews who escaped arrest often lacked the money to buy even a basic "A" ration.[80] Thousands of Jews and Romanies were interned, often under starvation conditions, and deported east to concentration camps.[81]

French scientists of nutrition in the mid-twentieth century combined a quantitative caloric measure, derived from German energetics, and racial-hygienic standard. Nutrition studies during the war acknowledged the importance of minimum energy intake; one study compared children's diets to a "strict vital minimum" caloric measure.[82] Other values were at stake as well, derived from a racial-hygienic tradition. Physiologist Lucie Randoin was one of the most visible nutrition experts of the late Third Republic and Vichy periods and was exemplary in that regard. Randoin posited three laws of human nutrition: the first set a "vital minimum" level of calories, following nineteenth-century energetic studies. Once "the necessary minimum of energetic substances" is met, she claimed, "the question of quantity is secondary." More important, for Randoin, was to achieve a "balanced" range of minerals and vitamins in accordance with racial and cultural norms.[83] "The role of diet," she wrote, "is to ensure perpetuity of the race, without any sort of degeneration."[84]

A select group of experts wielded exceptional influence on the design and implementation of state rationing from its inception to its end in 1949. In September 1940 the medical wing of the Academy of Sciences established a Scientific Consulting Committee on Provisioning, charged with "submitting recommendations that the Academy will issue to the French government, to doctors and to all those in charge of planning family or group diets."[85] One year later, the committee became a section of the Central Bureau of Research at the Ministry of Provisioning. The Ministry of Provisioning required its director of consumption and the director of food industries to attend each of the committee's meetings. The National Hygiene Institute and the Ministry of Health sent representatives.[86] An executive decree ordered the committee to "make suggestions in line with current necessities, on the subject of civilian population rationing and the preservation of public health."[87]

The committee's "duty to public authorities" was "to underline the important effects of a defective diet on the nation's general health."[88] The group formulated scientific opinions on the level of current monthly rations and "the distribution of rations across various categories of the population, by age and by category of workers."[89] Committee member Pierre Chouard was so implicated in rationing policy that he became one of its principal architects. Chouard, a botanist and professor at the National Conservatory of Arts and Crafts (CNAM), took a full-time position at the Ministry of Provisioning and remained in its employ until the end of the decade.[90] Chouard also composed *Dietary Resources in Times of Restriction*, one of a dozen popular guidebooks and cookbooks published to help people navigate the ration regime.[91] Dr. Charles Richet, an influential medical voice in the committee, was the son of a celebrated physiologist and a professor in his own right at the Paris Faculty of Medicine. In 1944 he was detained as a Resister in the Buchenwald concentration camp. In addition to Chouard and Richet, the committee's roster included four additional doctors, a physiologist, a member of the Superior Council on Hygiene, and two military intendance officers.[92]

In its public reports, the Ministry of Provisioning emphasized consumer choice, and dietary balance. The ministry assured the nation that food shortages would not have lasting health effects. Proper dietary balance, officials claimed, compensated and overcame the nation's state of penury. "Despite the undeniable insufficiency of our resources, [efforts and studies] are underway to provide vitamin supplements to children, to palliate albumin deficiencies and to educate the public with a few simple rules. . . . Thanks to this, we can look to the future without worry."[93] In the ministry's public declarations, science pointed the nation beyond the pains of hunger.

The ministry cited Lucie Randoin's theory of diet and racial health to reassure the population. In public speeches, Randoin promised that despite deeply reduced diets, French citizens could still maintain their health. A small amount of food could provide the caloric "necessary minimum." "Those who consume relatively little food—as is the case for many people today—can live just as healthily as those who wish to eat in great quantities."[94] Randoin again emphasized the importance of dietary balance; this was the task of scientifically informed, resourceful housewives.[95]

Yet the Scientific Consulting Committee at this same ministry estimated that "almost 100% of [French people] have undergone more or less severe weight loss" in 1942.[96] Around Paris, Drs. Huber, Collesson, and Rouech determined that "children are hungry." They cited data gathered by the chief doctor at the Lycée Hoche in Versailles in 1942 to the effect that 90 percent of students there had recently lost two to twelve pounds.[97] In summer of the same year the Union of Iron Miners sponsored a medical study of its 1,558 adherents in Lorraine. The researchers recorded "a general weight loss among the personnel, varying from fourteen to twenty-six pounds for an average weight of 140 pounds; an absolutely abnormal drop in arterial tension . . . symptoms of a weakening heart; the development of illnesses, particularly heart congestion, at a cataclysmic rate."[98] Members of the Scientific Committee warned in 1941 that only "a few hundred calories separate [current] restrictions from a semi-famine, and there is no guarantee that in future years restrictions will not be even more extensive."[99]

The National Hygiene Institute (INH) made important contributions to rationing data and policy during and after the occupation. Both during and after the war, the INH composed tables analyzing the "daily rate of rations" in major French cities and in sanatoriums, hospitals, and schools. The institute sent a monthly summary of its results to the Ministry of Health's regional offices across the country.[100] Researchers enumerated the calories, protein, fats, carbohydrates, sugars, and vitamins that an average resident absorbed in a given period.[101] The INH focused on medical studies of population. It sought to measure the frequency of pathological symptoms of malnutrition and to gather social statistics on dietary habits. "In times of scarcity, a rational distribution [of food] and dietary supplements can [only] be established by monitoring dietary status with studies of nutrition."[102]

The Ministry of Health established the INH in November 1941 and gave it responsibility for all public health research in France. The institute received a budget of 15million francs in 1943, which INH director André Chevallier spread across a number of scientific centers. Chevallier raised an additional endowment from the American Rockefeller Foundation. With the latter's help he established a Center for Hygienic Studies in Marseilles, funded nutrition research at the private Foch Foundation in Paris, and awarded medical school research fellowships.[103] The INH adopted a program of research specifically aimed to provide data to economic planners.

The INH's Center for Hygienic Studies in Marseilles produced some of the most prolific social statistical data. Between spring 1941 and winter 1946, the center carried out a "Study on the State of Nutrition of the Population of Marseilles" that resulted in hundreds of detailed family surveys.[104] Professional survey takers traveled across the city and enrolled families from all origins and walks of life. Participants weighed, measured, and recorded their daily diet, and also submitted to complete medical and hygienic testing.

A medical examiner tested participants for all possible symptoms of malnutrition and measured their "state of development." Inspectors described their subjects' cleanliness, family history, propensity for alcohol and tobacco consumption, and general mental and physical quality.[105] The Valente family, for example, had recently emigrated from Italy and depended on the mother's small salary from a local soap factory. Inspectors who visited the family in 1941 noted a caloric deficiency in widow Valente's four children. By contrast, little Roger Lévy, son of Jewish Algerian parents, received a high grade of "perfect subject—normal." The Guareschi family lived in a three-room apartment and "cooperated well" with survey inspectors. Father François, age thirty-two and a driver by trade, was in very good health despite a "clear riboflavin deficiency" and daily consumption of two liters of wine.[106]

On the whole, the Center for Hygienic Studies recorded an average daily caloric consumption of 30 percent less than the 2,400 recommended by the League of Nations. The INH directors effaced such embarrassing observations by hiding them in the inner sections of their reports. Over successive drafts of one such report, negative data on the French population's diet were shifted from the front page to the caption of one of its graphic tables.[107] Caloric intake did not return to normal until two or three years following the Liberation.[108] Still in 1945 Marseilles residents consumed 5 percent fewer calories than Parisians and 30 percent fewer than rural inhabitants; even larger gaps appeared in vitamin and nutrient intake.[109] Celebrated Resistance fighter Raymond Aubrac warned General De Gaulle that the city's situation required serious assistance: "People are dying in Marseilles, dying of hunger, full stop."[110]

THE VITAL MINIMUM WAGE

"A wartime economy," wrote François Perroux, "reinforces the method of a 'vital minimum.' . . . This method reverses the logic of purchasing power. It takes as given that the total economy must function to ensure the elementary and fundamental needs of all."[111] The 1941 Labor Charter [Charte du Travail] promised a vital minimum wage to all French workers. With this charter, the Maréchal Pétain became the first French head of state to institute a legal national minimum wage.[112] "The essential [goal] is to assure the minimum means of existence for the greatest number of families," pronounced Secretary of Labor René Belin in September 1940.[113]

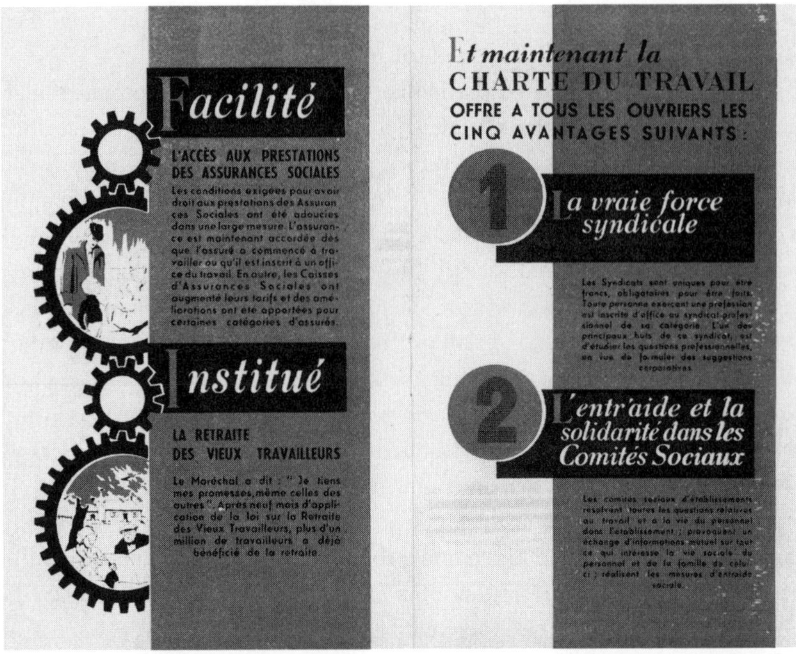

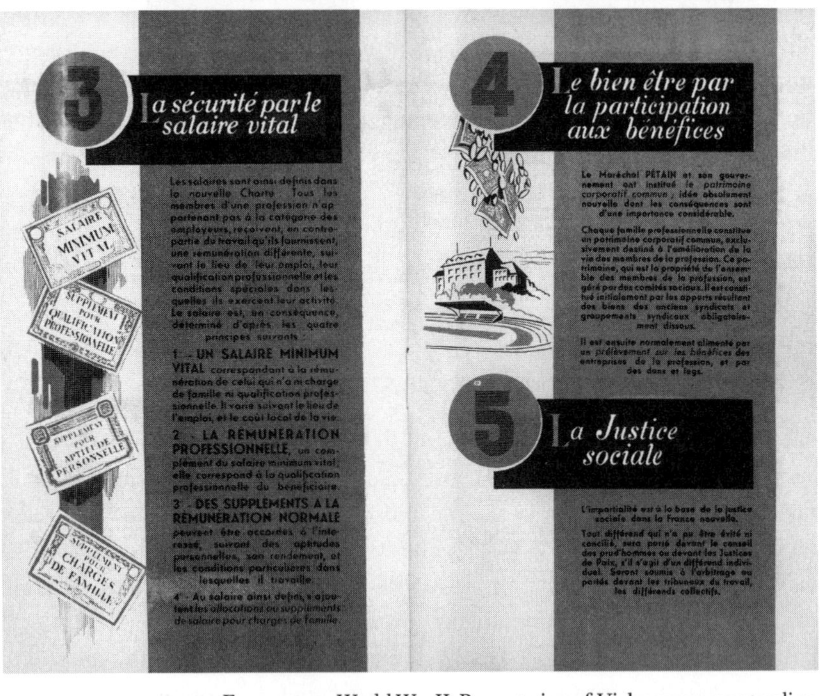

FIGURES 10 AND 11. France, 1941. World War II. Presentation of Vichy government policy: Labor Charter, minimum living wage and interest. © Roger-Viollet / The Image Works.

The Second World War and German occupation compounded the economic effects of the Great Depression, locking the entire population in a struggle to provide for basic needs. A flood of 8 to 10 million refugees fled the advancing Germany army in June 1940, abandoning food-producing areas in the north.[114] Big city dwellers bore the brunt of suffering; in 1941 a daily average of forty-one thousand Parisians relied on soup kitchens to eat.[115] During the hard winter of 1941–42, bread rations in Marseilles fell below the three-hundred-gram level (last seen during the Siege of Paris in 1871), meat consumption fell by 20 percent, and alcohol by 60 percent.[116] Finding food was a public obsession.[117] German requisitions gravely hobbled other areas of economic activity, including housing construction and nonmilitary industries.[118] Overall, workers' purchasing power fell somewhere between 30 percent and 40 percent over the course of the war.[119]

Even before the *Journal Officiel* published the Labor Charter, various committees were hard at work creating statistics on family budgets. The Vichy Council on Work drew from long-standing trade union practices of statistical collection. The council was filled with social Catholics and converted trade unionists. Trade unions were now banned, but other institutions took over their statistical project and gathered data on wages and prices. Vichy officials attempted to co-opt this tradition, historically associated with Socialist politics, within its new structure of professional corporations. The state offered an ambitious promise of a national minimum wage, based on scientifically established human needs.

Gaston Prache, secretary of the National Group of Consumer Cooperatives, estimated a "strict minimum of expenses, the vital minimum, indispensable for each of us." Prache, a former Communist and long-standing leader of the cooperative movement, was brought into the Vichy circle by René Belin. Prache sat on the National Council and the Council on Work and presided over the new state-sponsored cooperative system.[120] Prache's tables listed food, rent, heating, clothing, maintenance, and transportation costs for a family of three.[121] Former leaders of the outlawed Confédération Générale du Travail, in the guise of a "Committee on Economic and Syndical Studies," also presented a survey on "the minimum composition of the worker's budget."[122] Vichy appointed a Commission on Wages specifically to "establish budgets, for different geographic zones, corresponding to the physiological needs of an unmarried adult."[123]

The Council on Work called for "a vast local, departmental and regional survey . . . [that would] permit the National Commission to . . . set immediately usable bases for the minimum wage."[124] Local prefects, to whom the Ministry of Labor delegated authority to set regional wage standards, established statistics on the local cost of living and calculated typical low-income expenses. In the Loiret, a committee of experts, war veterans, merchants, and farm cooperatives added up costs for a number of consumer items.[125] War veterans and police officials in Clermont-Ferrand likewise surveyed expenses of Michelin factory workers and derived a minimum budget of 3,500 francs per month.[126]

Many workers argued that the minimum standard had to reflect "real," not ideal, conditions. As a research branch of the Ministry of Work argued, "If one wishes to give the vital minimum wage an operating definition that would allow legislators' intentions to be realized, to the extent that they are realizable, it is necessary to analyze the actual state of consumer goods on the French market."[127] "Given these conditions, what must one understand by vital minimum wage? It cannot be . . . the fictional purchase, at controlled prices, of the products in a 'menu' that is strictly indispensable for physiological needs. It is a notorious fact that it would be impossible for workers to find the necessary quantities on the officially sanctioned market."[128]

The vital minimum followed in a long tradition of budgetary wages, designed to provide for a basket of goods. A state-appointed wage commission recognized that "when the vital minimum is considered to be a measure of the local cost of living, it does not introduce anything new to a mechanism which, even before the war, linked wages to the price of consumer goods."[129] As a soap factory owner put it in 1942, "The fact is, when we pay men, we are thinking about their needs."[130] Vichy did not claim to originate this link; instead, officials boasted that they had introduced a "new notion: that of a lower limit."[131]

While the largest French families did see a rise in family allowances during the occupation, the vital minimum remained largely theoretical. Vichy authorities, under pressure to supply labor to the Germans, allowed the vital minimum wage to fall precipitously during their tenure. The occupying German administration blocked wage increases from 1942 onward.[132] Family support formed a cornerstone of Vichy politics and propaganda.[133] Yet the minimum wage remained as it was in 1917, a measure destined for single male workers.

Despite the protests of family associations, the official base wage covered only a bachelor's needs. Minister René Belin defined the vital minimum as "a minimum below which the effective wage of an unqualified adult worker should not fall."[134] Women's and apprentices' wages were calculated as a diminished percentage of the base measure.[135] Later, when working out a model budget for the vital minimum wage, officials attempted to reconcile the government's family rhetoric and its individualist wage standard. Commissioners at the Center for Wage Studies in 1944 "consider[ed] that an unmarried adult's budget should be taken as the basis for the remuneration of workers with family expenses, when setting out minimum needs."[136]

Ultimately the government withdrew from the promise of a guaranteed standard of living. René Belin was relieved of his post at the Labor Ministry, and the government moved markedly away from welfarism.[137] The vital minimum clause remained formally in place, but official press bulletins released in 1943 indicate that Vichy, now under direct German influence, had repudiated this policy. One bulletin condemned the vital minimum as a "brutal measure" which would cause wage hierarchies to collapse and stimulate price inflation.[138] Another announced that "the notion of a minimum wage has replaced that of a vital minimum."[139]

"If it is not possible to provide this country's workers the minimum that they demand, we fear that our members will no longer have faith in a reform that promised everything and delivered nothing."[140] The Departmental Union of Workers and Employers Unions in the Maine et Loire region warned ministry officials in 1943 that their promises had worn thin. Minimum wage legislation had little effect on French workers' ability to support their families. "Public health is in jeopardy. Children are not developing at the rhythm prescribed by growth charts. . . . Adults' health is even more alarming. . . . The worker faces a dilemma: either die of hunger or sink into gangsterism."[141] Scarcity of basic goods and tight food rationing obviated the promise of a vital minimum.

A contradiction stunted the promise of a scientific social state. The makers of Vichy social policy looked to science for a stable scale of value and for a guide to social organization. Real values would be found not in the marketplace but in human nature itself. As engineer and social reformer Jean Coutrot wrote in the 1930s, the aim of social science should be the "rational organization of human inequality."[142] Sociologists, biologists, psychologists, economists, and engineers would all contribute to a new understanding of human values. The human sciences would at last rise to the technical, rational standard of the physical sciences. Together they would establish the basis for an organic social harmony and a "science of man."[143]

Yet "authentic" needs were nearly unattainable under conditions of extreme scarcity. The promise of well-being contrasted starkly with real shortages and tight budgets. French workers' material wherewithal appeared to fail any acceptable standard. How, then, could one pretend to quantify an impossible level of well-being? Did an ideal minimum measure have any place in a world of hunger and low wages? "As soon as the vital minimum is defined in terms of physiological needs . . . [meaning] a given weight of products necessary for life, this minimum can only be guaranteed to all members of a community if a corresponding amount of total resources allow it. However equitable, it can only distribute available resources."[144] A scientific standard of biological health appeared incompatible with limited supplies. Yet with the Labor Charter's clause on the vital minimum, the government promised just such a measure.

THE SCIENCE OF MAN AFTER 1945

State economic planning in occupied France cast a shadow long after the war, as historians well have noted.[145] Vichy's policy-oriented scientific institutions also continued their work, and even expanded, in new forms. Some of the most striking continuities before and after the Liberation were in the field of economic policy. The *économie dirigée* called for by reformers in the 1930s, erratically implemented during the occupation, came to fruition after 1945. The Fourth Republic carried out Vichy's largely unfulfilled promise for a national need-based minimum wage.

After the Liberation the current National Institute of Demographic Studies (INED) grew out of the ashes of the defunct French Foundation for the Study of Human Problems, under the guidance of former Vichy functionary and demographer Alfred Sauvy. One of the foundation's most ambitious surveys, on housing needs and aspirations, was completed and published by the INED after the war.[146] The National Hygiene Institute remained in place until in 1964 it became the more global National Institute of Health and Medical Research (INSERM).[147] State-sponsored research organs proliferated, capped by a greatly empowered National Center for Scientific Research (CNRS).

Before and after the Liberation, François Perroux's economic and political commentaries remained strikingly consistent. Although after 1944 his publications were framed by quotations from William Beveridge rather than Alexis Carrel, their content almost exactly echoed that of his wartime pamphlets. In 1945, as in 1938 and 1943, he excoriated the values of liberal capitalism. The value of "human life" must replace the "logic of purchasing power."[148] The French state must continue to redistribute goods "according to the urgency and objective hierarchy of needs."[149]

Consumption remained at the center of Perroux's economic theory. "Economic power," he asserted in 1945, "also means the power to consume."[150] Consumption involved not only economics but the entire range of human sciences, "the physical-psychic domain."[151] Perroux and the other editors of the new journal *Economie appliquée* identified a fundamental challenge for postwar economists: "the problem of consumption," to match "a given good to a given need."[152] Once again, social policy should be set in response to scientific measures of human need. Economists had to undertake the "task . . . to learn how the wage earner transforms his purchasing power into real consumption, to compare his consumption to his needs, and to gather a set of useful indicators for the economy of reconstruction, of France today."[153]

In October 1944 François Perroux founded the National Institute for the Science of Applied Economics (INSEA). This institute quickly became the central node for economic thought in postwar France.[154] John Maynard Keynes was "one of its first patrons," and it attracted visits by many of the major economists of the time, including Joan Robinson, Paul Streeten, Thomas Balogh, and Joseph Schumpeter. Michael Kalecki held a permanent position there.[155] The institute's journal, *Economie appliquée*, gained a wide readership. Perroux's influence was rewarded in 1955 with the chair of "Analysis of Economic and Social Facts" at the Collège de France. He took on directorship of a new research center, the Institute for Economic and Social Study and Development, in 1960, the same year that he was named to the state Economic and Social Council.[156] Perroux thus regained a position comparable to that he had held in 1941–43, a social-scientific expert in the wings of political power.[157]

State-enforced rationing ended only in 1949. The Scientific Consulting Committee continued to evaluate and formulate rationing plans; indeed, its influence seems to have increased. The committee set a stringent state policy of minimal rationing in prisons and internment camps for Jews and Romanies.[158] Public pressure for a fair system of wages and distribution grew louder as channels for political expression reopened. Communists, technocrats, social Catholics, and local administrators involved in building the Fourth Republic all agreed that the state should guarantee a just economy of human needs.[159]

CHAPTER 8

Human Persons

In mid-May 1950, the Subcommission on the Nondietary Part of the Budget-Type held a particularly fractious sitting. André Baupaume of the CFTC union (Confédération Française des Travailleurs Chrétiens) submitted that sanitary facilities in a "minimum dwelling" should include a sink with running water and a drain. The assembled union leaders, employers, and state officials had all concurred on gas, electricity, and a toilet for every floor, without debate. Not so for the sink. "A long discussion opened," report the meeting records. "The CGT [Confédération Générale du Travail], CGT-FO [CGT-Force Ouvrière] and the CFTC consider that a source of running water inside the dwelling is indispensable," a part of a worker's vital minimum.[1] Representatives from all the country's major unions, despite wrenching ideological conflicts in the political sphere, coalesced in unity around the water tap.

Not the employers. "The CNPF [Conseil National du Patronat Français] and the CGA [Confédération Générale de l'Agriculture] think that currently, the question of running water inside the dwelling only can be considered desirable. The majority of dwellings do not contain running water; a sink, in general, is located one to a floor."[2] Indeed, only 18 percent of rural homes boasted of running water in the early 1940s.[3] Statistics gathered in the Seine region by tax authorities in 1939 and 1940 suggest that the ratio of apartments without bathrooms to those with a bathroom was at least ten to one.[4] This situation certainly did not improve during the war and occupation, given that 20 percent of France's housing stock had been destroyed between 1939 and 1945.[5] How could one propose running water as a vital, indispensable good when most French people didn't even have it?

On the other hand, "M. Bauhaud (CFTC) noted that this refusal of basic and indispensable comforts, particularly running water within the dwelling, would compress the budget for rent so low that it would come to less than 1% of [a worker's] total resources, which is completely insufficient."[6] Housing authorities agreed that workers already allocated far too small a portion of their pay to their landlords.[7] A century of hygienic reformers had campaigned for access to running water. If they failed to account for the sink, the Nondietary Subcommission risked contravening the advice of decades of experts. In the end, the bosses gave in, and commissioners included one square meter of water per month, for urban residents only, in their model worker's budget.[8] In an echo of nineteenth-century respiration studies and social housing design, commissioners agreed to set the minimal dwelling's living space at fourteen square meters per person.[9] Tap water and living space thus became part of a new national minimum wage, the *Salaire minimum interprofessional garanti*.

Citizens of the postwar French welfare state negotiated their status through questions of consumption. Material goods such as tap water came to represent and embody contested social hierarchies. Welfare and wages were measured by standards of living, the goods required for a socially acceptable quality of life. As an ever-greater number of citizens entered the economy of wage work and mass consumption, that distinction became an important class marker. Class conflict was objectified in debates over water, toilets, and basic goods.

This is a story of class politics, science, and consumption entangling in the postwar French welfare state. In spring 1950, the French minister of labor asked a group of union leaders, employers, and experts to design a new national minimum wage. They were charged with defining a "budget-type," an exemplary monthly budget for the lowest-paid French worker. In setting its budget, the wage commission drew a baseline portrait of the minimal consumer. In debates over everyday goods, commissioners grappled with the frustrations and desires of postwar French citizens.

Until well into the "glorious thirty" years that followed World War II, Europe's major economic concern was insufficiency, not excess. The postwar European consumer society was born of scarcity. American historians have accustomed us to seeking the symptoms of a postwar consumer society in abundance, overproduction, and ostentatious self-display. The "American standard of living" long stood for the comfort and security enjoyed by an "average man" such as a worker in Henry Ford's five-dollar-a-day plant.[10] The European experience, however, offers a very different perspective. Into the 1950s, social and economic life was shaped by conditions of penury.[11]

The postwar French state dealt with poverty and scarcity by setting standards of living. Like the Vichy government before it, the Fourth Republic turned to neutral, scientific measures in order to sideline class confrontation. Social actors were obliged to translate their demands into a nonpartisan budgetary language. This arrangement, in theory, should have resulted in a transparent and democratic polity. Leaders in the new republic designed its laws to reflect the values of "democratization, national solidarity, and rationalization."[12] Scientific experts and associations had a strong, often decisive, voice in wage and welfare policy. Yet from the very opening of wage talks in 1950, it became clear that standards of living would not easily produce solidarity or consensus.

The story of standards of living, needs, rights, and science is tied to the fate of the modern welfare state. According to the "rules of coordination" of welfare capitalism, decent wages purchased workers' "commitment, loyalty, and effort."[13] The French welfare system relied on a transparent and consensual distinction between needs and luxuries, welfare and the market. Yet no one could agree on the list of basic goods and services that the state should guarantee.

Science promised to establish a clear measure of human need. Twentieth-century French governments employed scores of dietitians, sociologists, anthropologists, and doctors to establish a vital minimum, an objective consumer standard. They collected mountains of statistics on people's consumption and income, their diets, housing conditions, and fitness. By the postwar period, however, it had become clear that no single scientific standard could encompass the rapidly changing lifestyle of the average European consumer.[14] Political discord between workers and employers, each wielding competing sets of data, rendered it impossible for science to play a mediating role. Expert knowledge promised to depoliticize the social question. Again and again, it failed to do so.

INCOMPRESSIBLE NEEDS AND THE SMIG

The passage of minimum wage legislation in 1950 led to one of the odder chapters in the history of French bureaucratic committees. Paul Bacon, minister of labor, convened a High Commission on Collective Bargaining in April of that year to measure a model worker's budget. Over thirty members strong, the meeting assembled major leaders of all three unions, representatives of large and small employers, artisans, managers, and family associations, as well as a number of state bureaucrats.[15] Over the following two months, these men

FIGURE 12. Paris, France: September 5, 1952. First session of the Superior Committee of Collective Agreements at the Ministry of Employment. © Roger-Viollet / The Image Works.

hashed out the principles and the practice of a new minimum standard of living. No point proved too fine for debate, from theoretical reflections on science and measurement to the durability of underwear.

The minimum wage commission offers an exemplary moment in postwar French consumer society. First, it enacted a political contest between unions and employers, mediated by the state and family associations. Second, as it called upon expert testimony, the commission participated in the emergence of an empirical, policy-oriented postwar social science. Third, the commission's work reflected French citizens' everyday struggle to reconcile scarce resources and expanding consumer desires.

Scientific experts, unions, and employers struggled to define lower limits for the normal consumer. Were human needs social, psychological, or physiological? Did the vital minimum refer to an absolute limit of bodily survival, to the energy required for productive work, or to customary norms and aspirations? Who was this minimal consumer: man or woman, married or single, with or without a family? Each of those questions required consensus if a stable standard of living was to emerge. The commissioners' inability to agree on a basic unit of analysis contributed strongly to their ultimate failure.

Paul Bacon introduced the commission to its task by recalling the Fourth Republic Constitution's promise to "guarantee the necessary conditions for the development of the individual and the family."[16] But how would they define and quantify these "necessary conditions"? Henri Raynaud, CGT secretary, opened discussion and laid out the terms on which his union was prepared to negotiate. "M. Raynaud judges that the only problem that the commission can study at this time is that of a budget-type. . . . The CGT considers that this problem cannot be properly addressed unless one adopts the notion of a vital minimum."[17]

"What is the vital minimum?" the CGT newspaper *Le Peuple* asked in 1946. "For the biologist it is the sum of calories necessary for the individual's physical life. The sociologist and the economist translate it into francs according to the prices of basic goods. . . . For us, in addition, it is a certain margin of well-being that does not condemn the least-paid worker to a purely animal or vegetative life."[18] After the Liberation, the CGT and other unions recycled a premise put forth by the Vichy regime, that wages should be set with respect to a scientific measure of human need. Vichy's Labor Charter, promulgated in 1941, guaranteed a vital minimum to all paid workers. Although the Labor Charter's demise followed that of the occupied government, the concept of a vital minimum remained present in the popular press and among union members. The term itself did not appear in the 1950 wage law, but it became the de facto guiding concept for the High Commission's negotiations.

Marcel Meunier, president of the Union of Metals and Mining Industries and member of the employers' confederation, emitted an immediate objection. The wage law of 1950, he noted, said nothing about vital needs, only minimum needs. "One must conclude that the minimum wage is not necessarily tied to this complex notion. We must distinguish what is minimum from what is vital."[19] Meunier demanded that the commission "judge the minimum or the vital character of each line [of the model budget]. Dietary needs do correspond to the meaning of vital; on the other hand, one can debate the inclusion of metro tickets in a budget of national scope."[20]

Everyone agreed that the vital minimum, whatever it was, must lie beyond the frontier between life and death. "The biological minimum cannot be the minimum below which one cannot go without punishment of death; in that case, one could argue that survivors of Auschwitz and Mathausen had received a minimum."[21] In the immediate postwar period images of starving men, women and children from cities and camps were still raw. On all sides of the social and scientific debate, everyone acknowledged that such terrible wasting could not form the basis for a social standard.

Having set aside the threshold of death, experts left a field open to divergent opinions and measures. No single point emerged as an obvious limit to human existence in 1950s France. Some doubted whether needs could be quantified at all. The Vichy government, two years after having erected a legal vital minimum in its Labor Charter, repudiated the notion as a "brutal" and "unsolvable" measure.[22] Others found the "solution" to minimum standards elusive. Did needs have fixed and stable limits, or did they increase and expand continually? By 1950 some argued for a "psycho-sociological minimum . . . which recognizes a worker's right to participate in the material and cultural progress of civilization."[23] Could a minimum measure progress? Or did such an approach result in oxymoron, as when one union proposed the neologism "average vital minimum"?[24]

Beyond the commission's doors, standards of living were fluctuating painfully. The post-Liberation years veered agonizingly between penury and the promise of satisfaction. By 1947 it became clear that the war's end had tightened, not eased, food rationing. Due to an unsuccessful harvest and vacillating economic policy, bread rations fell to two hundred grams in the fall of that year—lower than they had ever been during the occupation.[25] Many French were shocked to discover that the rations American soldiers accorded to German prisoners of war were higher than those set by their own government for its citizens.[26]

National productivity regained its prewar level in 1947, but standards of living did not follow. "At each stage, inflation lowered morale, unraveled plans and projections, reintroduced inequality and reactivated social unrest."[27] At the Liberation, French residents lived at approximately 60 percent of their prewar comfort level, according to one historian's calculations. Living standards rose to 85 percent of that level by summer 1945, but quickly turned backward and "neared 50 percent of that same level in May 1947."[28] Given that the state controlled both wages and prices, provincial opinion blamed government incompetence for the people's economic woes.[29] Everywhere citizens suspected that peasants, Americans, ministry apparatchiks, or black-marketeers were withholding supplies from the market, which everyone thought should be abundant.[30]

Even as material conditions degenerated, consumer hopes expanded. The Fourth Republic Constitution, enacted in 1947, guaranteed "the conditions necessary for the development of individuals and families." It promised "to all, notably to children, mothers and elderly workers, the protection of health, material security, rest and leisure."[31] This right was realized in part with the passage of national Social Security in 1945. A greater range of goods came

FIGURE 13. Paris, France: July 1948. Inauguration of a self-service store. © LAPI / Roger-Viollet / The Image Works.

within eyesight in shops and markets, only to deceive eager purchasers with their stratospheric price tags. The allure of American abundance enveloped U.S. forces as they traversed the country and wafted across the Atlantic in films and advertisements.

French citizens' struggle to satisfy their basic needs played out in movements for higher wages. Resistance leaders had considered a "vital minimum wage" as part of their plans for a post-Liberation France as early as 1942. A special commission on the minimum wage was established in the section assigned to economic, financial, and social questions.[32] The National Resistance Council released a plan for liberated France in March 1944, which included "the guarantee of a wage and payment level insuring that each worker and his family will have security, dignity and the possibility of a fully human life."[33]

Frustration over the gap between goods and aspirations fed massive wildcat strikes in 1947. Striking public workers kicked off one of the most widespread and violent social movements in modern France. These actions grew from a grassroots level. They confronted leaders' productivist politics with urgent consumer demands.[34] Local unions and their members pressured national leadership into a campaign for a state-mandated minimum wage. Resolutions calling for a vital wage were issued by numerous federations.[35] In response to their constituents' urging, the CGT, the CFTC, and the CGT-FO unions all publicly claimed the vital minimum as their own.

The social movements around 1947 put workers in a position of force. For a short moment workers and unions held the upper hand over employers in the French balance of political power. In fall 1946 the government ceded to public employees' pressure and dictated that a vital minimum serve as the basis for state wages. Three years later every wage earner in France, public or private, gained legal right to the *Salaire minimum interprofessionel garanti.* The law of February 11, 1950, ended wartime wage controls and instituted a state-enforced minimum wage. Better known as the SMIG, this new minimum wage immediately impacted the least-paid industries: textiles, clothing, metals, agriculture, chemicals, paper, construction, and lumber.[36] By 1951, 65 percent of French workers lived on a SMIG wage.[37]

HUMAN PERSONS

The SMIG commission of unions, employers, and experts thus set out to design a model consumer budget. Some of the same men sitting at Paul Bacon's table had been there three years before, negotiating a standard "vital minimum" for setting public workers' wages.[38] Although the material conditions of 1950 were far improved over those in the difficult year 1947, SMIG commissioners chalked their work onto that previous effort. The 1947 law on public functionaries offered the following rule of measurement: "One should interpret the vital minimum as the sum below which a human person's individual and social needs, considered to be basic [*élémentaire*] and incompressible, can no longer be satisfied."[39] But who, in 1950 France, was this "human person," and what did he or she require?

Pro-family groups argued that the standard "human person" ought to represent a family unit. A representative of the National Union of Family Associations (UNAF), Louis Leroy, "asked that the commission first establish a family budget, then deduct an individual budget from it."[40] For decades pronatalist organizations had campaigned for a wage standard based on family size. These associations earned strong support across the political spectrum, as the nation succumbed to fear of demographic decline. Thanks to their influence, welfare policy since the interwar period largely focused on subsidies for family support.[41] Wages, however, were firmly excluded from family considerations.[42]

Despite the very strong influence of family politics in mid-twentieth century France, family support never broke the legal barrier between wages and welfare. Family allowances were established in the early part of the century as charitable supplements and always remained in the legal realm of welfare. Although these allowances were mandated by the state beginning in the 1930s, they were not counted as part of a worker's salary for the purposes of unemployment insurance or retirement. Union leaders in the SMIG commission battled to retain this separation. Wages were for workers; welfare was for families. Employers had no reason to disagree with the unions on this score. They had long preferred to keep family allowances separate from wages; by raising the former, they avoided costly increases in the latter.[43]

Union leaders in the SMIG commission were not about to accept that wages be confused with welfare payments. Upon hearing Leroy's proposal union leaders jumped to their feet. "M. Delamarre (CGT-FO) cannot accept M. Leroy's method. . . . The family problem must be considered in the legislative context, after the *salaire garanti* is set." Gustave Salmon of the CFTC "reminded [the commission] that family supplements present different problems than the question of wages."[44] In other words, the unions fought to keep wages separate from the question of family welfare. The SMIG's model consumer would thus be an unmarried, childless worker.

Was this unmarried worker a woman or a man? Before the war, standards of living were exclusively masculine. The first minimum wage laws in the early twentieth century covered only women workers, in an attempt to reduce cut-rate competition. Women, as supplemental workers, were seen as a threat to the male wage scale.[45] Vichy's 1941 Labor Charter, on the other hand, guaranteed a minimum salary only to men. Resistance leaders in London chose to include both men and women when planning for the postwar society. Planners called for a uniform, ungendered minimum wage, in recognition of "women's dignity" and to prevent "competition from cheap labor": "The essential needs of each individual depend on common factors like local conditions, social milieu and the real cost of products. . . . On the basis of . . . the equality of needs, the minimum wage should be uniform."[46] Following the lead of Resistance planners, the Fourth Republic eliminated gender from its wage laws. The SMIG did not distinguish between male and female wage earners. A decree in 1950 confirmed that the SMIG applied to "workers of both sexes . . . at least 18 years of age."[47] Postwar law created a new category of worker: not family, man, or woman, but a "human person."

The apparent universality of this moniker, however, quickly disappeared from view. As the "human person" took on concrete professional and regional characteristics, it once again essentially excluded women from consideration. Unions and family associations objected to universal standards for all people. The United Nations Food and Agriculture Organization, for example, designed a global caloric measure; the League of Nations also had defined a universal caloric minimum in 1936 for a sedentary individual. The 1947 commission decided that it could not countenance such abstraction. Universal standards, commissioners argued, risked mischaracterizing the real needs of French workers. By the time it had finished, it defined its "human person" with surprising specificity: "The case [we have] chosen is that of an unmarried light worker [elsewhere called a broom sweeper], housed in an unfurnished apartment in Paris, and who cooks and cleans."[48]

The SMIG commission formally ratified the definition of vital minimum contained in the public functionary law, including its reference to the basic needs of a generic "human person."[49] As negotiations proceeded and numbers were crunched, however, this abstract unit disappeared. CGT delegate Roger Pascre deemed that the reference to "social needs" in the meaning of minimum vital "implies the notion of work"; that is, social identity as a worker was embedded in the very definition of the vital minimum.[50] In subcommission, the participants felt free to embroider. At first, they directed the dietary part of their model budget to an "ordinary manual laborer [*manoeuvre ordinaire*], meaning a broom-sweeper, a worker who is not physically fatigued."[51] At some later point in the commission's work the human broom sweeper became an even more particular figure: an unskilled metals worker residing in the Paris region.[52]

Employers, naturally, objected to the "human person's" concretization in the form of a member of a powerful workers' constituency, at home in the most expensive region in France. "We continue to think that the Commission took the wrong turn when it abandoned the definition of individual and social needs of a human person, considered as basic and incompressible, to evaluate [instead] the needs of a manual laborer in the Paris region." To add insult to injury, the other commission members, against employers' veto, had "added on elements which are neither vital nor minimum."[53] The commissioners' inability to agree on a basic unit of analysis contributed strongly to their ultimate failure.

Even this semi-concrete measure, the manual laborer, came in for mockery and criticism. "We won't tarry today on the question of which underpants a 'human person' who is dressed [*culotté*] by M. Barjonet [CGT] should wear in order to secure his basic and . . . incompressible needs," joked a broadsheet. The satirical "Society for Economic and Documentary Studies" offered a "Little Sketch of the 'Human Person' at the Dinner Table," and criticized his meager dietary rations.[54]

Officials at the Ministry of Labor's Statistical Division struggled to compare the commission's model worker to a real social group. Asked by the ministry to verify the SMIG commission's budget, the director of the National Institute of Statistics and Economic Studies (INSEE) threw up his hands. The "human person" could not be found: "It is appropriate to point out that an unmarried manual laborer living alone, preparing his own food, and cleaning his home appears very infrequently and therefore cannot be the subject of a scientific [statistical] study."[55]

Given that the commission's "typical" subject was in fact an exotic rarity, its entire calculation lost any meaning. How could one claim to measure the budget of a nearly nonexistent entity? "Therefore, for almost every consumer item, any decision concerning quantity . . . must be largely arbitrary, and the statistical data that we currently dispose of can only reduce a small portion of that uncertainty."[56] The Statistical Division office consulted its data and ruled that the "human person" did not exist.

A different challenge to the commission's model, launched a few years after passage of the SMIG, contested the assumption that all workers were in fact "human persons." A Doctor F. Bardin, medical control officer, wrote in to protest the government's enforcement of minimum wage laws in the department of the Sarthe. "In the Sarthe there is a non-negligible proportion of psychologically retarded people. This fact is explained by the alcoholism rife in the region. These psychologically retarded people have a small output [at work] which does not allow them to be paid a wage equal to the SMIG." Bardin, despite his self-described "social spirit," made a plea to the central office to intervene in support of the low wages in his area. "The needs [of these workers] are incontestably lower than those of a normal wage-earner or worker."[57] The Ministry of Labor agreed, at least in part. It allowed employers to identify no more than 10 percent of their workers with "reduced physical aptitudes" and to reduce their wages by one tenth.[58] It appears that the "human person" also required an unspecified quantity of intelligence and physical strength.

If the "human person's" identity could not be justified scientifically, commissioners sought out some aspect of the vital minimum that they could calculate with mathematical precision. They found it in nutrition. Dietary measures provided a seemingly incontestable standard. "The vital need par excellence is nutrition."[59] Wage commissions in 1947 and in 1950 divided their model budgets into dietary and nondietary sections. "One must make a critical distinction between a global vital minimum that . . . includes all sorts of expenses for a decent life, and a biological vital minimum that only deals with diet. It is essential to note that only the biological minimum can be the subject of rational and scientific calculation."[60] The dietary category would include all items that boasted a scientifically endorsed limit; the other would contain everything else. The first would be physiological, the second a social measure of need.

Dietary Budgets

To its regret, the commission soon discovered that food was no scientific haven. Nutrition science itself roiled with debate over how to measure needs. In vain, the SMIG commission and all of the unions called expert upon expert. Commission staff compiled a collection of all dietary studies conducted in France since 1940.[61] Commissioners took testimony from Drs. Lucie Randoin and Georges Duchêne of the Scientific Society for Dietary Hygiene, and Dr. Jean Trémolières of the National Hygiene Institute (INH). Yet they could not obtain a consistent answer.

Employers' representatives sought to revive the League of Nations' 1935 dietary standard for sedentary people. "As far as the food budget is concerned, when dealing with basic and incompressible needs, we believe that the only incontestable scientific basis [for this measure] is the figure of 2400 calories established by the Hygienic Commission at the League of Nations."[62] Unions and family associations insisted that a supplement be added to any caloric standard to account for energy expended in work. "[The League of Nations'] standard suffices for a sedentary man or woman, but is insufficient for the worker."[63] Lucie Randoin, in a 1946 report on rationing, distinguished between a "vital minimum" of 2,200 calories and an "optimal quantity" of 2,800 calories.[64] The dietary subcommission ultimately adopted a figure of 2,890 daily calories, which naturally employers found excessive. The latter pointed in protest to a study by the Foundation for the Study of Human Problems, which found a shoemaker living happily on 2,000–2,200 calories per day.[65]

But calories could not be the only element in a nutritional standard. If the commission adopted a purely caloric measure, employers might be tempted to fill a sack of potatoes and call it a daily minimum. In an earlier period, Vichy officials had latched onto the ideal of a balanced diet in order to downplay caloric shortages. The CGT, by contrast, pushed for dietary balance in order to justify more and varied food expenses. "If we did not take account of qualitative factors we would not be able to respond to our opponents' argument that we could always obtain the necessary minimum calories by eating bread, pasta or potatoes in large quantities and still live."[66] Although no employer took the foolish step of publicly suggesting such a diet, a purely caloric standard did leave open that possibility.

As soon as one left the realm of calories, however, hard and fast rules were ever more difficult to come by. Even the biological minimum could not escape the contaminating influence of social habits and expectations. To prescribe a balanced diet meant itemizing each menu and meal component. On what basis should one choose between bread and pasta? Nutrition experts could not agree. Both the CGT and the CFTC advanced scientifically endorsed "theoretical rations," model menus for light workers. Yet commissioners already in 1947 had rejected these as impractical and impossible to obtain.[67]

Jean Trémolières, director of nutrition at the National Hygiene Institute since its foundation in 1941, testified before the dietary subcommission at its second meeting in 1953. Trémolières argued passionately that nutrition science must have an empirical sociological basis. Nutrition should measure not calories but "the influence of social milieu on consumption."[68] "For many physiologists," Trémolières wrote in 1952, "the notion of 'need' is absolute; but [in fact] it can only ever refer to a man for whom we have defined his physiological characteristics, his work and his lifestyle as closely as possible."[69] Large-scale social surveys should observe "rations actually consumed" on average by different social groups.[70] "[Nutrition] standards must be grounded only in the actual consumption of real, active groups in good health."[71]

But as soon as one applied Trémolières' prescription for a social, empirical standard, it became clear that there could be no standard at all. Take, for example, milk, fruits, and vegetables. Everyone agreed that the average French family consumed terribly insufficient amounts. Should the commission retain an ideal nutritional standard, one far greater than most people's actual habits? Dr. Duchêne of the Scientific Society for Rational Hygiene said yes. He argued that a large allotment of fruits and vegetables in the SMIG budget would "improve . . . the worker's dietary situation."[72] But Trémolières vehemently differed. As he warned the commission in 1953, "each time that an attempt has been made to normalize and modify dietary habits in the name of laboratory experiments, it has encountered serious difficulties."[73] Trémolières thought that the model budget ought to conform to French people's eating habits by reducing fruits and increasing fats.[74]

Some participants questioned the implications of basing minimum standards on social statistics. Roger Pascre of the CGT worried that when one "begins with the observation of [dietary] habits and defines a model ration on that basis, one finds oneself in a vicious circle. If we only deal with actual consumption, we end up maintaining the current situation in place, even though nothing proves that it is satisfactory."[75] Pascre's concern echoed the judgment of some nutrition experts. Sometimes, one scientist objected, needs expand to meet new norms. Children drank more milk in 1950 than before the war, because of new calcium standards. Which to choose, expert standards or real habits? The two could become locked in a "vicious circle of norms and surveys."[76] If dietary norms took current consumption as their basis, they would do no more than reproduce the already existing state of scarcity.

Wage commissioners rallied to Trémolières' contention that the only way to reach a "determination of standards [is] to observe the spontaneous consumption of groups of workers."[77] The 1947 commission chose to create its model menu on the basis of social statistics. "Given the impossibility of precisely measuring the food products currently available, the commission was led to examine actual consumption, as described by the results of studies in large cities by the National Hygiene Institute [INH]."[78] It seems to have gone unnoticed that INH nutrition surveys of Paris and Marseille had warned that urban residents suffered from a severely insufficient diet.[79] State employee's minimum wages were set by social statistics, which themselves were judged inadequate when compared with theoretical standards.

A more obvious obstacle prevented the use of theoretical rations for setting the vital minimum: scarcity. This proved especially true in 1947, when food resources were stretched thinner than at almost any point in modern French history. A model measure containing even just 750 grams of bread had little meaning next to an official ration of one-third that size. Although conditions improved by 1950, they were still far from ideal. "Within the quantitative limits of theoretical need . . . the commission seeks [to define] a ration that satisfies needs and takes account of goods actually found on the market."[80] Market dynamics, too, came to intrude upon the security of scientific dietary measures. The gap between needs and wants intruded on the commissioners' model diet.

At the end of the day, the SMIG commission gathered every available estimate of dietary need and simply took an average of the whole bunch. These included, in utter confusion: studies by the National Hygiene Institute of "insufficient" urban diets, the previous (1947) commission's dietary estimate, government statistics on current agricultural output, and the Planning Commissariat's predictions for future production.[81] No one could seriously claim that the resulting number in any way lived up to the promise of a "rational and scientific" measure. They could only bow to Trémolières' contention that "the definition of a minimum standard is scientifically impossible."[82]

Nondietary Budgets

The promise of a rational, universal dietary standard was obscured by social realities. Yet this should have been the easy, consensual part of the SMIG negotiations. All other kinds of human need, by comparison, were obscured by a fog of speculation. "If the evaluation of a biological minimum is grounded in scientific bases that allow us to narrow the problem down and to make a precise estimate, the situation is completely different for the other items included in a vital minimum."[83] As the satirists above had noted, underpants were much more difficult to prescribe and quantify than dinners.

Commissioners hashed out the components of a decent, but not excessive, life. The 1947 Statute on Public Functionaries, to which all parties agreed to adhere, stipulated that the vital minimum include social as well as bodily needs. As we saw above, Marcel Meunier of the employers' association made his opening gambit by denying that any expense other than food should qualify as "vital." The unions accused employers of squeezing the minimum to mere subsistence. "For the bosses, the vital minimum is guaranteed as soon as the worker gets a paycheck that allows him not to die of hunger, go naked or sleep under bridges."[84] The CNPF's own internal documents, however, show that employers never expected the vital minimum to contain only edible items. Its model budget contained the same products as those found in the unions' studies.[85]

The employers' hardline stance might have been motivated by their handicapped position. The 1947 strikes convinced public authorities that citizens' consumer demands had to be placated, at least to some extent. Large employers bore the brunt of Fourth Republic social measures, including the cost of family allowances and a large part of Social Security.[86] In addition, Minister Paul Bacon set up the SMIG commission such that family associations, represented by the UNAF, cast a tie-breaking vote. Where consumer welfare was concerned, the UNAF regularly sided with union demands. Employers were fighting a coalition between a Fordist high-wage ideal, citizens' desires for an abundant consumer society, and a well-established family welfare tradition. They lost on every significant issue.

Employers sought to prescribe and reduce every piece of the model worker's consumption. Leaders of the CGT, by contrast, resisted attempts to enumerate social expenses. Workers, they argued, should be free to choose which goods to purchase and how often. In this the unionists revealed surprising fidelity to an expansive consumer society, even to an American model of consumer empowerment. They refused to limit the horizon of workers' lifestyles and comfort. Perhaps, too, the union sought to maintain a class-based consumer identity. If statistical averages governed workers' clothing and leisure activities, consumption might lose its function as a class marker.

Social-scientific experts offered their services as mediators in this politically charged debate. A handful of sociologists, economists, and statisticians undertook research specifically designed to aid in the official measure of a nondietary vital minimum. In many ways they followed a tradition of state-sponsored social-scientific research put into place in occupied France. "Scientists of man" already had gained a privileged role in setting the Vichy regime's wage- and food-rationing policies.[87] Experts proved no less important in the post-Liberation government, which began ambitious programs of national accounting and economic planning.[88] Social scientists had good reason, therefore, to expect that the wage commission might seek their aid.

As the unions and employers sat down to work, the experts followed course. In June 1951 the National Center for Scientific Research (CNRS) gathered a group of important figures in nutrition, sociology, and statistics, with the common task of defining a minimum budget.[89] Representatives from major state research institutions—the National Hygiene Institute, National Institute of Statistics and Economic Studies, experts from the Ministry of Agriculture and Ministry of Provisioning—were present.

Most prominent in this group was the young sociologist Paul-Henry Chombart de Lauwe. Of Catholic and military origins, Chombart became an ethnologist under the tutelage of Marcel Mauss.[90] At first he had planned to study impoverished populations in Asia; the Second World War would forever turn his attention and ethnological skills to poor French at home.[91] At age twenty-seven, Chombart became one of the key figures at the Vichy-era Uriage school for future Catholic leaders. He came in close contact with economist François Perroux and the Dominican social researchers at the Economy and Humanism group, both important influences on early Vichy social policy.[92] Like many other social Catholics, Chombart's loyalty to the Vichy regime weakened after 1942, and he was eager to aid the new republic after the Liberation.

Chombart identified a fundamental challenge awaiting all attempts to define a minimum budget in 1950s France. The danger, he recognized, lay in the fault line between "the need for something that one already has, and the need for something that one does not yet possess."[93] More simply put, the gap between "needs and aspirations." Some needs could be measured, physiologically or by averaging consumers' purchasing habits. But aspirations, worries, and dreams about the future could only be accessed psychologically. If one examined only current consumption, one would miss out on a whole range of needs, needs projecting into the future.[94]

Chombart's work combined an economics of scarcity with sociologist Maurice Halbwachs's vision of consumer abundance. Chombart shared Halbwachs's commitment to progress, growth, and expanding comfort for the working classes. His postwar research on Paris and its environs drew heavily from Halbwachs and garnered officials' attention.[95] Chombart signed research contracts with several ministries and formed a special policy research group within the Center for Sociological Study (CES). He saw his group as an intermediary between the government and the governed. If the state, as he believed, existed "to satisfy the people's needs," then his role was to define those needs that required satisfaction.[96]

France's leaders, Chombart complained, were incapable of grasping the people's needs and desires. He criticized measures like the SMIG for sticking too closely to the level of "obligatory subsistence needs [*besoins-obligation relatifs à la subsistance*]." Such a vital minimum would do no more than "allow one category of the population to survive by providing the ruling class with the labor that it requires." Instead, he set out an alternative program for policy and research: "for everyone, in the context of a harmonious *cité*, to seek the means to satisfy all of the human needs."[97]

Chombart thought even workers and their unions incapable of moving beyond a narrow vision of consumption. "Workers themselves have defended a vital minimum based on satisfying 'basic' needs." The unions' model budgets all reflected workers' current habits, not what they should or would like to consume. Workers and their unions had thereby unknowingly assimilated and reproduced "a representation of the ruling classes."[98] Certainly no Marxist himself, Chombart argued that leaders ought to envision needs stripped of all reference to a particular class.

If the working classes were too "absorbed by day-to-day struggles" to evaluate their own needs, sociologists could provide the "necessary analyses."[99]

Chombart dedicated his long research career to working-class consumption and need. His group at the CES gathered a staggering amount of data on Parisians' everyday habits: their work, commuting, and housework schedules; the size, cost, and arrangement of their homes; their monthly incomes and expenses; their everyday meals and how they gathered for Sunday dinner. In the interstices of the data, Chombart sought to uncover worries and desires. Meat, for example, meant much more than a supply of protein: if workers ate more than they "needed," it was because they associated this food with freedom from "hunger and misery . . . the fear of falling [down the social ladder], of not 'holding up.'"[100] Housing, too, could reveal desires and aspirations: in the suburbs, workers hoped to buy their own little home; in the city, they aspired to a larger apartment. Chombart composed "indices of satisfaction and desire for change."[101]

Indices of future desire, however, proved both unwieldy and inconvenient for the SMIG commission. It was already under an impossible pressure to bring its budget up to current standards, much less potential aspirations. "Recourse to theoretical speculations" about the social and psychological bases of need, complained the INSEE statistical institute's director in 1951, was "unproductive."[102] After fourteen years of lobbying to make the sociologist's voice on wages and urbanism heard by deciders, Chombart eventually retired his policy group out of frustration and turned to more academic pursuits.[103] Policy makers were bereft of a simple social-scientific solution. They would need to draw on their own resources to reconcile needs and aspirations, scarcity and abundance.

Members of the nondietary section thus used their own judgment to navigate the world of everyday needs. Commissioners found themselves steeped in the banal details of common goods. What should workers buy, and how often? Products that workers should own were not necessarily those that they did own or wished to. "If an adult's need for a fixed number of calories or vitamins daily has a scientific basis, on what grounds could we decide that he or she needs an overcoat every four years instead of a heavy coat every six years or a raincoat every winter?"[104] Here, even more than in debates over diet, consumer desires clashed with the real availability of goods on the market. If scientific measures offered no solution, the commission simply fought it out.

Clothing raised a heated debate. How thick must a worker's jacket be, and how threadbare should he wear it? What about a tie or scarf? Some suggested that the commission measure clothes' durability and include this depreciation in the budget. They called for statistical studies on average replacement time for common household goods. The CGT, on the other hand, argued that the commission should not presume to design a worker's wardrobe or any such personal expense.[105]

Leisure, vacations, wine, and coffee presented the stickiest points of contention. Commissioners ran into trouble on the latter points because both drinks had been shown to lack nutritional value (though the caloric content of wine still remained under debate). Yet alongside bread, wine and coffee were the very staples of an actual French diet. The 1947 commission decided to "account for the French nation's habits" with a small quantity of coffee.[106] On the other hand, it excluded tobacco from its calculations. After debate, the commission resolved to include wine in its "ration-type" on the basis of a "psychological argument." It thereby recognized the "stimulating effects of this national drink."[107] Experts agreed: "For most households, at least for men, wine remains a vital necessity."[108] The SMIG commission fully endorsed the human person's need for wine and counted seventeen liters of wine monthly, nearly a bottle a day, for every French wage earner.[109]

"The vital minimum must account not only for purely physiological needs, but also for a psychological minimum (different kinds of distraction, reading, sports, etc.), by which men distinguish themselves from beasts."[110] On that basis, both the CFTC and the CGT asserted that "leisure constitutes a basic and incompressible expense." "The worker has a right to the resources that permit him to have a minimum of intellectual culture and leisure."[111] Unions inherited the proposition that vacation is an economic right from the Popular Front, which instituted the first mandatory paid vacations in 1936.

Small business owners summarily rejected the notion of a minimum leisure standard. Big bosses were somewhat more accommodating, at least in spirit. The CNPF "recognizes the worker's need to rest in the countryside during his vacation, but cannot admit that such an expense be considered as a basic and incompressible need." Employers justified their apparent stinginess by suggesting that the SMIG should be considered only a temporary status. "As soon as the laborer improves his situation, normally after two years, it will be possible for him to budget for these expenses."[112] This optimism did not appear in sync with the overwhelming proportion—two-thirds of the French salaried workforce—about to benefit from a SMIG level of pay.

Economic planners in the Resistance government also had argued for leisure and cultural needs. The national minimum wage, as foreseen in 1942, would allow workers "to have access to a small margin of resources to pay for so-called 'little extras' (tobacco, theater, cinema, different outings) and also a small reserve . . . to buy the accessories of life which everyone else [has] but which currently remain out of reach, if only a simple radio."[113] The 1947 commission on the vital minimum included a sum more than three times that of rent for vacations, "sports, leisure, and culture."[114] "If keeping current with political or other news and continuing one's education through reading is not absolutely indispensable in order to live, commission members nevertheless considered this to be a social need."[115]

Radios, vacations, newspapers, and sinks represented some of the panoply of goods expanding before consumers' eyes in the immediate postwar period. Yet these products, along with a full diet and durable wardrobe, remained beyond many would-be purchasers' reach. Measurers of the vital minimum were in a bind. Products that French people desired, those scientists deemed they should have, and those within reach on a worker's pay were dramatically disparate. Ultimately the unions and family associations combined to push through a generous and unspecified standard. The commission "arbitrarily decided to include either realistic, or desirable, consumer goods instead of scientifically [statistically] determined consumer habits."[116] The SMIG model budget thus included a hefty sum for vacation (one and one-half times the monthly rent) and a margin of 10 percent of total expenses for "leisure, culture, and [union] membership fees."[117]

AN IMPOSSIBLE STANDARD

Through the 1950s Labor officials were annoyed by a constant flow of letters from around the country asking to be informed of the *salaire minimum vital*. Small-time workers requested that the minister tell them how much they should be earning. The ministry prepared a form letter to send to these inquirers: "I have the honor of informing you that the expression 'vital minimum', although commonly used, is improper. The Ministry of Labor and Social Security has never set such a minimum."[118]

Ultimately, the SMIG men failed in their impossible task. Nearly from the outset of negotiations, employers refused to cooperate with the basic principle of measuring a minimum. The CGT also eventually turned its back on the vital minimum. The union had enthusiastically promoted it beginning in 1946; one article in *Le Peuple* even went so far as to claim paternity of the concept.[119] By the close of the SMIG talks, however, the union publicly repudiated the notion as "an insult to the working class." Where they once celebrated a guaranteed minimum standard, the pages of *Le Peuple* now carried biting opposition. "It took time and compromises for our phantom government to decide that the

FIGURE 14. Paris, France: 1973. Poster for Lutte ouvrière (Trotskyist political party), campaign to raise the SMIG to 1,500 francs. © Roger-Viollet / The Image Works.

minimum garanti would be a minimum of misery."[120] Local CFTC members complained bitterly that their leaders had shown "excessive moderation" and "a lack of combativity," in the SMIG negotiation.[121]

The SMIG commission failed to reach an agreement on the essential values and norms of the postwar French consumer society. Fourth Republic wage laws had attempted to circumvent class conflict with a scientific, consensual standard of living, the vital minimum. Yet standards of living failed to produce consensus. Ultimately, political forces—the relative power of unions and employers, the influence of para-political associations, the strength of state enforcement—substituted for neutral scientific measures. The rational, transparent, universal citizen of the postwar republic dissolved in the face of competing interests and political strife. Class struggle played out in negotiations over how to define a minimal French consumer.

In the process, social actors revised their prewar views of consumption. Unions campaigned for a society of mass abundance. Communists, despite a reputation for anticonsumerism, praised consumer choice and product variety. Family associations and scientific experts largely supported the union position, by tying needs to consumer choices and desires. Employers, on the defensive and seeking to reduce costs, found themselves fighting against a Fordist high-wage ideal. Workers, experts, and employers put forth opposing visions of the postwar consumer society.

From a sociological perspective, the actors seated around the ministry's table had clear class affiliations; they were quite aware of their opposing interests and identities. The content of their debates suggests a more complex picture. They were charged with creating a new social hierarchy based on consumption and material goods. In this case the language of class combined economic and political rights with gender and social stratification. Class distinctions appeared in the form of what Pierre Bourdieu called *habitus*, in assumptions concerning consumers' tastes and needs.[122]

When the assembled social representatives failed to agree on a minimum measure of need, the government imposed an arbitrary and nonbinding solution. After three months of fruitless haggling, the SMIG commission sent a draft budget without majority approval to the minister of labor. The minister himself, forced to define a wage standard without the commission's unanimous guidance, soon repudiated the vital minimum. Officials returned to a simple cost-of-living index, without making any claims for its normative value. The government could not guarantee a minimum standard based on either fixed scientific norms or expanding consumer needs and desires. Within two years new strikes would force a renegotiation, and the cycle began again. In a society ruled by scarcity and lured by the promise of growth, the science of human needs could not provide a stable ground for economic policy.

Finally, in 1970, President Georges Pompidou severed the French minimum wage from consumer statistics entirely. A new measure, the *Salaire minimum de croissance* (SMIC), was indexed instead to GDP, national productivity.[123] The more liberal state of the 1970s overturned the assumption, once widely shared, that life could be subject to scientific measurement and regulation. The new regime of wages and welfare swept aside the vital minimum.

In the 1970s quantitative standards of need lost their normative power. "Quality of life" became a political platform: John F. Kennedy used the term in his 1963 State of the Union speech, and European leaders soon followed suit. French prime minister Georges Pompidou even briefly installed a "Ministry of Quality of Life" in his 1973 cabinet. "Quality of life" committees emerged in contexts as diverse as the Organization for Economic Cooperation and Development (OECD) and urban planning groups. Even the field of accounting took a new interest in this period in the "motivational complexity of a person," "in order that the systems and their methods of use might be redesigned, so as to enhance the well-being of person and organization."[124] As a result of this movement, tables of costs and benefits gave way to more complex considerations of institutional culture, leadership, and personal preference. Social needs gave way to individual preferences and rational choice.

CHAPTER 9

Need, Nature, and Society

> Locke produced a defense of private property based on the natural right of man to that with which he has mixed his own labor, and many thousands of people believed and repeated this, when it must have been obvious to everybody that those who most often and fully mixed their labor with the earth were those who had no property, and when the very marks and strains of the mixing were in effect a definition of being propertyless. . . . Once we begin to speak of men mixing their labor with the earth, we are in a whole world of new relations between man and nature, and to separate natural history from social history becomes extremely problematic.[1]

This book is about the politics of life under a modern European form of welfare capitalism. Social scientists, at the inception of their disciplines, strove to resolve the contradictions apparent in post-Revolutionary society. One of the thorniest concerned inequalities of wealth and well-being. Natural law justified the unequal distribution of private property. Yet other natural principles, like Lavoisier's chemical physiology, revealed a fundamental imbalance. Those who labored most had little food; those who labored least had plenty. Workers who bore the "marks and strains" of transforming natural resources into goods had little access to their products. Science formalized and problematized this contradiction. This was a moment at which natural history was purposefully inserted into social history.

Ramachandra Guha has argued that "there was no environmentalism before industrialization."[2] One might make a parallel claim here: there were no human needs, in a scientific or political sense, before the age of wage labor. "Needs" expressed an economic order—the distribution of limited resources, and a social order—the organization of human differences. Needs became a problem precisely when both sides of that equation—the social and the material—were subject to choice (through political action) and transformation (through industrial activity). Wages, welfare, and social policy were all open, at least formally, to negotiation. Collective actions, not custom or divine right, determined the distribution of power and goods. Needs served to elaborate a science of the modern welfare state.

The binding of natural and social history appears even tighter when one considers it in context. The post-Revolutionary period experienced one of the greatest physiological transformations in human history. Demographic historian Robert Fogel calls it an era of "technophysio evolution." Intensive agricultural and industrial productivity accompanied unprecedented increases in body weight, height, and energy. Fogel contrasts the eighteenth century, when smaller bodies survived by expending less energy, with the modern era of increased consumption, larger bodies, and heavier labor.

"Subsistence," Fogel argues, came to mean something entirely new in the modern West.[3] People "subsisted" in the old regime, but with higher mortality and smaller size. "Individuals in the bottom 20 percent of the caloric distributions of France and England near the end of the eighteenth century lacked the energy for sustained work and were effectively excluded from the labor force"; those who did work were generally weak by modern standards.[4] In the modern period more food was made available, bodies grew larger and resisted disease more effectively, and people were more economically productive. This transformation spread in Western countries even to the lowest social classes due to public investments in health and welfare.[5] Fogel estimates that "the average efficiency of the human engine in Britain increased by about 53% between 1790 and 1980." Growth accelerated most rapidly in the twentieth century. That increase, he claims, accounted for enormous gains in productivity and fully half of British economic growth in the same time period.[6] Subsistence—need—became a scientific problem in the modern West at a time when the balance between food (and other goods), body size, and work underwent dramatic transformation.

The modern subsistence regime meant much more than a numerical increase in food supplies, heights, and weights. Subsistence has two sides. One is physical—providing energy, health, and strength; the other is social—providing for well-being, productivity, and reproduction. Measures of body size, for Fogel, stand in for much broader effects on productivity and quality of life. Technological change and physiological growth, he claims, are "synergistic, which means that the total effect is greater than the sum of its parts."[7] Stronger, healthier bodies could produce more and longer; higher productivity could lead, under the right circumstances, to a better quality of life. I would add a third element to this synergistic interaction, quite possibly with Fogel's approval: the rise of the modern welfare state. Western European states initiated public health programs. Just as importantly, states regulated the distribution of material wealth and well-being. The interaction between workers' bodies, industrial economies, and the welfare state creates what Fogel calls "physiological capital."[8] Physiological capital has a social as well as a biological content. The two sides of subsistence—nature and society—are, on closer examination, part of a single movement.

Scientists of human need spent a great deal of time trying to distinguish the "natural" from the "social" and to set one of those terms against the other. Either nature or society was believed to hold information about "real" needs. Scientists argued that one or the other ought to be eliminated from calculation.

In the late eighteenth and early nineteenth centuries, the "natural" appeared real. Men like Lavoisier, Condorcet, and Lagrange sought to regenerate French society with nature as their guide. Their vision of a homeostatic, balanced natural order offered a measure and a template for post-Revolutionary society. Real human nature could be uncovered in physiology, as in the chemical respiration experiments of Lavoisier, Dumas, and Boussingault. Or it could be sought among primitive peoples, those "majestic monuments to the origin of time" whose ways revealed a state of nature.[9] Gérando and Le Play guided their reform-minded readers to "penetrate" poor people's existence; a century later Vichy ethnologists repeated that exercise in the far reaches of rural France. For these men, grounding needs in nature ensured that social hierarchies would remain fixed and unchanging. Nature also served a more progressive social program. Even those with a Lamarckian, evolutive view of needs assumed the primacy of nature, whereby people's sensations, needs, and physical bodies would alter in response to changing environmental circumstances.

The turn to nature implied that society should bend to fit nature's mold. Lavoisier suggested, in vague terms, that physicians be empowered to set the distribution of wealth according to each worker's needs. Dumas, Boussingault, and Gasparin carried out a variation of this program through state policies on prison, school, and army dietaries. Many in the nineteenth century, including Dumas, Gérando, Auguste Comte, and the Saint Simonians, believed that a natural hierarchy of talents, abilities, and needs could best be accommodated by a scientific-capitalist social order. These men, and others including Le Play, explained and justified specific forms of social inequality on the basis of unequal natural talents and needs.

Around the turn of the twentieth century, roughly speaking, this dynamic reversed in France. As the SMIG commission's debates illustrate, by the mid-twentieth century the "real" was to be found in social statistics. "Real" consumption—recorded through social surveys—appeared a better guide to human needs than "theoretical" physiological measures of vitamins and calories. This shift resulted from a triple movement. First, social statistics were adopted by workers and trade unions early in the nineteenth century in opposition to the "natural" social hierarchies proposed by Gérando and Le Play. Second, the Third Republic established a language of citizenship, which prescribed needs according to one's utility to the state. Finally, welfare policy made the distribution of goods a question of social management rather than of physiology.

An analogous shift appears in nineteenth-century English public health, where medical prescriptions for basic needs—food, clothing, and heat—were replaced by a "fragmented and incoherent list of imperatives" that focused more on prescribing sanitation than on maintaining bodily health. "In famine-accommodating post-Malthusian political economy," writes Christopher Hamlin, "the question of biological needs could be avoided because there was no expectation that the natural systems in which humans existed would supply those necessaries."[10] Instead of measuring natural needs, English sanitarians focused on managing hygienic behavior. Needs left the purview of medicine for that of social engineering.

Emile Durkheim's theory of anomie suggested that social norms, not natural laws, grounded social order. Only a stable and authoritative society, he argued, could maintain consensual norms. At each level of the hierarchy, in such societies, people made lifestyle choices in accordance with a common sense of propriety. For each profession, norms dictated a minimum set of basic needs and a maximum level of acceptable luxury. This stability had been lost in industrial society, he lamented, due to the unpredictable movements of markets, goods, and people. Many social reformers shared this fear of urban consumption and its fleeting satisfactions. Durkheim's sociology of consumption is echoed in critiques of consumer society from the late nineteenth century to the present. Real needs—stable, hierarchical, embedded in authoritative social norms—were contrasted with desires, fleeting and unfixed. Social stability derived from social cohesion and organization, not from any natural referent.

The French debate on standards of living intensified at the turn of the twentieth century, with the inception of private and public welfare. Sociologists, economists, religious social reformers, and industrial employers grappled with how to reconcile wages and welfare, production and reproduction. The equation between needs and wages could not hold, and employers found themselves under attack by workers demanding cost-of-living raises. Through industrial employers' efforts, the French welfare system was designed to shield debates over needs from a broader social conflict over wages. Charitable family allowances were meant to cover social needs, which in principle freed the wage to be set by productivity and the market. In France, as elsewhere in Europe, needs were associated with realm of welfare, while desires and luxuries appeared to belong in the realm of the free market.[11]

Different metrics served competing visions of social needs. Should one measure real incomes or the relative distribution of wealth? Should one weigh the foods that a working family consumes or count the relative weights of food, clothing, and housing in a family's budget? Anthropometric measures of height, weight, and body mass or fertility and mortality rates? Psychological satisfaction and happiness?[12]

The same methodological question animated the postwar British Standards of Living debate. Historians, set off by a heated exchange in the late 1950s between R. M. Hartwell and Eric Hobsbawm, argued whether life got better or worse with industrialization. The liberal "optimist" school represented by Hartwell based its case principally on charts of rising real incomes. Hobsbawm accused these figures of obscuring an understanding of "real consumption" and high relative inequality.[13] "The historian," he charged, "forgets at his peril that the problem of the social impact of the industrial revolution is not whether men live by white or brown bread, no meat or roast beef. . . . It is also, that men do not live by bread alone."[14] In the background of the debate, as economic historian Stanley Engerman recalls, was a political contest between capitalist accumulation and Communist collectivization.[15]

Many economic historians since have offered data on wages, consumption, the distribution of goods within communities and families, and anthropometrics. Absolute economic measures tend to show improvement in consumption over time for all people; relative measures have been used to demonstrate the extent of exploitation and inequality. Anthropometric measures tend to fall in between: English (and American) workers' heights stagnated across the economic growth period of the mid-nineteenth century, then climbed markedly. Body height measures also reveal clear biological divisions along class and gender lines.[16] Over time, historians shifted emphasis from the question of "what happened to real wages in the Industrial Revolution?" to "what happened to the relationship between working and living conditions?"[17]

The English Standards of Living debate was not only about metrics. Through competing data historians set "quantity of life" against "quality of life," "standards of living" against "way of life."[18] Aspirations, social bonds, justice, and equity appeared as important as ounces of bread or nominal wages. Some economists and psychologists turned to a "happiness index" to track satisfaction as the primary indicator of well-being. Psychologists now admonish consumers to pursue experiences and relationships, rather than material possessions.[19] (Interestingly, such claims appear in an age when a declining industrial sector is dominated by the service industry.) Amartya Sen elaborated Hobsbawm's critique into a theory of standards of living centered on human diversity and accomplishment. Sen's method accounts for what people can do, rather than simply what goods or wages they collect. "What we can or cannot do, can or cannot achieve," he suggests, "do not depend just on our incomes but also on the variety of physical and social characteristics that affect our lives and make us what we are."[20] Rather than seeking a single standard by which to measure every individual's existence, economists inspired by Sen now tend to cluster many related factors such as education, literacy, mortality, morbidity, food, housing, and health care.[21]

For many twentieth-century socialists, "real" social needs could not be apprehended by any measure; they had yet to be created. German Bauhaus architect Walter Gropius used this argument to promote high-rise apartment blocks as a solution to the interwar housing shortage. It did not matter, he argued, what today's workers appeared to need. Architects and sociologists had to measure people's real "natural and sociological minimum requirements, unobscured by the veil of traditionally imagined historical needs."[22] The role of architecture, wrote a Czech modernist architect, was to "engender new functions and awaken new needs."[23] Russian Constructivists of the 1920s applied the same program in designing new objects—uniforms, reading rooms, kiosks—of everyday life.[24] Social theorist Herbert Marcuse accused the postwar consumer society of perpetuating false and "repressive needs," which stimulated a superficial, destructive satisfaction. "The most effective and enduring form of warfare against liberation," Marcuse warned, "is the implanting of material and intellectual needs that perpetuate obsolete forms of the struggle for existence."[25] For Marcuse, individuals could elaborate their "true needs" only under conditions of social liberation—meaning freedom from social constraint. "True needs" lay beyond the horizon of historical experience, in nonrepressive forms of social interaction and individual expression.

In the late twentieth century, nature re-entered the equation. The "limits to growth" era set human needs against natural resources. As in the eighteenth century, nature appeared to have ultimate authority over social organization. Nature presented an inexorable limit to expanding human needs. Western consumerism seemed likely to strip natural resources beyond a point of no return. The specter of scarcity and limits came as a shock to many Western consumers in the post-Fordist era. It seems, however, that the thirty-year postwar Golden Age stands as a historical exception rather than a rule. Even as political scientists in the 1980s declared the dawning of a "postmaterialist" era, the material conditions on earth experienced unprecedented change.[26] Both agriculture and industry intensified production and resource exploitation in what environmental historians call the "Great Acceleration."[27] Vastly expanding appetites of consumers at the global center led to the stripping of much of the land and resources of poorer producers at the periphery.[28] Now the spread of consumer needs across developing countries—India and China foremost—is blamed for impending ecological disaster.

Natural limits, alongside the decline of the welfare state, have forced Western consumers to reconsider our needs. In the early twenty-first century, needs again appear a pressing social and political question. What do we really need? Can we find a common answer to that question, and how are norms for satisfaction built and maintained? What are the limits to consumption; are they natural or social, and who decides? Can abundance be reached by all, or does the comfort of some depend necessarily on the privation of others? Can we reshape our satisfactions to fit the constraints of scarcity and limited resources?

To seek an answer to these questions is to challenge the historical opposition of nature and society. Philosopher Kate Soper eloquently suggests that " 'flourishing' is what we ought to be re-thinking in the light of our current and future resources; it is not an a priori given of human nature whose 'true' needs nature can be expected to fulfill."[29] She imagines an "alternative hedonism" that would recast "what is to count as the achievement of 'health' or 'sanity' or the 'enjoyable life.' "[30] This "erotics of consumption" would allow people to conceive of satisfactions grounded in their material environment and social interaction. Human needs, self-development, and self-fulfillment require that we reconcile our material and social existence.

ACKNOWLEDGMENTS

I had submitted a request through a computer terminal at the Bibliothèque Nationale for a box of "Rationing cards, World War II." An hour later I received one of those responses to which researchers resign themselves: the material is not appropriate for distribution. Access denied. Ah, well, I thought, and turned to the next book on my list. I had barely begun to hammer in the first few relevant quotes when an unknown face stopped at my numbered desk. My name is Madeleine Barnoud, she said, I am curator of special objects and I've decided it's a shame that we don't let people look at them. Thus began a most extraordinary adventure. She took me behind the librarians' desk, with its imported wood paneling, to the other half of the Bibliothèque. There, motorized tracks mounted on concrete walls ferried books across a semi-industrial workspace in shocking contrast to the luxury outside. She led me to an elevator and pressed the third-floor button. Somewhere between floors two and three the elevator stopped and the three librarians inside began to jump. Noting my surprise, they explained that this happened rather regularly and a good jostle should get it going again. Which it did. I spent the better part of an hour or two sifting through worn, half-used ration cards in Mme Barnoud's office. None of them made it into this book. However, none of the book would have been possible without untold hours of labor and kindness on the part of Mme Barnoud and her colleagues in France, Germany, and the United States.

The Interlibrary Loan staff at the University of California, Riverside—in particular, Janet L. Moores and Esther Arroyo—deserve a monument to their greatness. My thanks, also, to their counterparts at the University of Chicago. In France I would like to thank Guillaume Marchand at the Institut français d'architecture, Anne Kuhnmunch at the Confédération française des travailleurs archives, and the long-suffering staff at the Bibliothèque Nationale, Archives Nationales, Service historique de l'armée de terre, and the Centre des archives contemporaines.

I have had the great fortune to carry out this project with a non-minimal standard of living, thanks to the support of generous people and institutions: the Jacob A. Javits Fellowship program, the Eric Cochrane Traveling Fellowship, the University of Chicago Group on Modern France, the Institut d'Etudes Politiques de Paris, the Graham Architectural Foundation, the University of Chicago William Rainey Harper Fellowship, the Max Planck Institute for the History of Science, the Coordinating Council for Women in History, the Forum for the History of Human Sciences, the Department of History at the University of California, Riverside, and the UCR Center for Ideas and Society.

In the days when one drank to History on the deck of the Bateau Ivre or the John Hancock tower, I gathered many debts of friendship and wise council. My thanks to those who sustained me and prodded me forward: Michael Geyer, Leora Auslander, Robert Richards, Myles Jackson, Aaron Hill, Steve Sawyer, Mark Loeffler, Paul Cheney, Paul Dutton, Norton Wise, Ted Porter, Mary Terrall, Christopher Hamlin, Michelle Murphy, Frank Trentmann, Alexander Nützenadel, Christian Gerhardt, Anne Lhuissier, Emma Spary, Soraya Chadarevian, Wolfgang Schivelbusch, Hannah Landecker, Adriana Craciun, and John Tresch. Yossef Saadia, the Przeworskis, and the Paris dissertation group (Ben Kafka, Julie Coe, Andrew Jainchill, Rebecca Manley, Charly Coleman, and Anoush Terjanian) made intellectual life beautiful.

Ursula Klein and Hans-Jörg Rheinberger took me in and supported my stay at the magical center of intellectual effervescence that is the Max Planck Institute for the History of Science. I am grateful to them and to my colleagues on the fifth floor of the old Czech embassy for an unforgettable year of excitement and exchange. In Los Angeles, Eileen Gibson Funke and my fellows at the Writers Junction provided the best company a writer could ask for. Jim Clark and Peter Baldwin both read and commented on entire drafts of my manuscript out of pure and extraordinary generosity; they helped me to understand what a book looks like. Three anonymous readers at the University of Chicago Press took the manuscript apart and gave me brilliant suggestions for reassembly. Karen Merikangas Darling made this book real. I am enormously grateful for her steady hand and perceptive advice.

Finally, those debts for which words do not suffice. I owe my intellectual life to Jan E. Goldstein, Ken Alder, and Ann Goldberg. I can only hope to pass on to others the friendship and support that these three have given me over the years. My dear friends and family filled to overflow that most profound of my human needs: joy. (I mean you, Lazare Simmons-Saadia!)

To Pierre Birnbaum, who years ago warned that this project was far too expansive for a first major research undertaking: perhaps you were right.

NOTES

CHAPTER ONE

1. Outward Bound, "Dan's Environmental Corner — Simplicity, What do you really need?"

2. Warren McLaren, "Stuff: How Much Do You Really Need?"

3. Stephanie Rosenbloom, "But Will It Make You Happy?"

4. Frederick Cooper and Randall M. Packard, "Introduction," 7–9.

5. Kate Soper, "Conceptualizing Needs in the Context of Consumer Politics," 358. Soper's insightful philosophical analyses of the politics of need have been invaluable to this work.

6. The blue, white, and red flag was adopted by the Assemblée Constituante on October 21, 1790.

7. Antoine Laurent Lavoisier and Armand Séguin, "Premier mémoire sur la respiration des animaux," 691.

8. Frederick L. Holmes, *Lavoisier and the Chemistry of Life*, 452.

9. Lavoisier and Séguin, "Premier mémoire sur la respiration des animaux," 691.

10. Lavoisier and Séguin, "Premier mémoire sur la respiration des animaux," 691.

11. Holmes, *Lavoisier and the Chemistry of Life*, 450.

12. M. Norton Wise, "Mediations," 217.

13. Bernadette Bensaude-Vincent, "The Balance," 228.

14. Lisa Dicaprio, "Women Workers, State-Sponsored Work, and the Right to Subsistence during the French Revolution," 524.

15. Lavoisier and Séguin, "Premier mémoire sur la respiration des animaux," 698.

16. Lavoisier and Séguin, "Premier mémoire sur la respiration des animaux," 698.

17. Larry Glickman, *The Living Wage*, 62.

18. Sidney Webb and Beatrice Potter Webb, *A History of Trade Unionism*, 325–26. Glickman cites Lloyd Jones as the origin of the term "living wage." Glickman, *The Living Wage*, 62.

19. Nikolas Rose, *The Politics of Life Itself*, 352.

20. Keith Tribe, "Continental Political Economy," 161. See also Keith Tribe, *Strategies of Economic Order*, 72.

21. Pierre Larousse, "Subsistance," 1175; Pierre.-J. Proudhon, "Qu'est-ce que la Propriété?" 343.

22. Moishe Postone, *Time, Labor, and Social Domination*, 161.

23. David Harvey, *The Limits to Capital*, 49.

24. David Ricardo, *On the Principles of Political Economy and Taxation*, 93.

25. Maxine Berg, *The Machinery Question and the Making of Political Economy, 1815–1848*, 44–45.

26. Giorgio Agamben, *Homo Sacer*, 88.

27. On the imbrication of these two structures, see Bruno Latour, *Reassembling the Social*, 328.

28. François Vatin, *Le travail, économie et physique, 1780–1830*, 127; Anson Rabinbach, *The Human Motor*, 402; Margaret Schabas, *The Natural Origins of Economics*, 231; Timothy Mitchell, *Carbon Democracy*, 288; John Tresch, *Romantic Machines*, 17; Ludmilla Jordanova, "Interrogating the Concept of Reproduction in the Eighteenth Century," 369–86; Michelle Murphy, "Distributed Reproduction," 21–38.

29. Karl Marx, *Capital*, 612.

30. John Shovlin, "The Cultural Politics of Luxury in Eighteenth-Century France," 605.

31. Jean-Claude Perrot, *Une histoire intellectuelle de l'économie politique, XVIIe–XVIIIe siècle*, 185.

32. William H. Sewell, Jr., *Work and Revolution in France*, 217.

33. "25–29 février 1848. Déclaration du gouvernement provisoire relative aux ouvriers," in J. B. Duvergier, *Collection complète des lois, ordonnances, règlemens et avis du Conseil d'Etat*, 59.

34. Alain Bayet, "L'accroissement spectaculaire des salaires et leur pouvoir d'achat," 156.

35. Olivier Marchand and Claude Thélot, *Le travail en France, 1800–2000*, 51.

36. Robert Castel, *Les métamorphoses de la question sociale*, 352.

37. Giovanna Procacci, *Gouverner la misère*, 164.

38. Susan Pedersen, *Family, Dependence, and the Origins of the Welfare State*, 17.

39. Joan Wallach Scott, "'L'ouvrière! Mot impie, sordide,'" 143.

40. Londa L. Schiebinger, *Nature's Body*, xii.

41. James Vernon, *Hunger*, 273.

42. William Coleman, *Death Is a Social Disease*, xv; William Reddy, *The Rise of Market Culture*, 140.

43. Sheila Jasanoff, "Idioms of Co-production," 6.

CHAPTER TWO

1. Charles de Remusat, *Mémoires de ma vie*, 177.

2. André Jardin and André Tudesq, *Restoration and Reaction, 1815–1848*, 292–95.

3. Jean Boulaine and Jean-Paul Legros, "Gasparin," 57.

4. He owed his first administrative appointment in 1830 to the ascendance of his dear friend François Guizot, interior minister of the new July Monarchy.

5. Louis Blanc, *Révolution française*, 253. Blanc, a socialist and harsh critic of Gasparin, quite likely embellished in recording this pronouncement. However, it is true to Gasparin's political spirit at the time.

6. Claude Latta, "Lyon 1834," 23.

7. William H. Sewell, Jr., *Work and Revolution in France*, 209–17; Yves Breton and Michel Lutfalla, eds., *L'économie politique en France au XIXe siècle*, 559.

8. Boulaine and Legros, "Gasparin," 58. My thanks to Jan Goldstein for pointing me toward Gasparin's political history!

9. F. W. J. McCosh, *Boussingault*, 63.

10. Mary S. Morgan, "Experimental Farming and Ricardo's Political Arithmetic of Distribution," 48; Lluis Argemí i d'Abadal, "Agriculture, Agronomy, and Political Economy," 449–78. See also François Dagognet, *Des révolutions vertes*, 105–9.

11. Dagognet, *Des révolutions vertes*.

12. Morgan, "Experimental Farming and Ricardo's Political Arithmetic of Distribution," 5.

13. Peter McPhee, *The Politics of Rural Life*, 36, 40.

14. Nicolas Bourguinat, *Les grains du désordre*, 27.

15. David H. Pinkney, *Decisive Years in France, 1840–1847*, 49.

16. William Reddy, *The Rise of Market Culture*, 148.

17. Dagognet, *Des révolutions vertes*, 143.

18. Dagognet, *Des révolutions vertes*, 142.

19. John Pickstone, "How Might We Map the Cultural Fields of Science?" 356–57.

20. Dana Simmons, "Waste Not, Want Not," 73–98.

21. Jean-Antoine Chaptal, *Chimie appliquée à l'agriculture*, 298.

22. Jean Baptiste Boussingault, "Recherches expérimentales sur le développement de la graisse pendant l'alimentation des animaux," 423, 440.

23. Jean Baptiste Boussingault, "Analyses comparées des alimens consommés et des produits rendus par une vache laitière," 114.

24. Boussingault, "Analyses comparées des alimens consommés et des produits rendu par une vache laitière, 113–44.

25. Bernadette Bensaude-Vincent, "The Balance," 224.

26. Frederick L. Holmes, *Lavoisier and the Chemistry of Life*, 450.

27. M. Norton Wise, "Mediations," 207–58.

28. Wise, "Mediations," 222, citing Lavoisier, "Prix proposé pour l'Académie des sciences pour l'année 1794" (1792), in *Oeuvres* 6:34.

29. Lissa Roberts, "Condillac, Lavoisier, and the Instrumentalization of Science," 258, 267.

30. Antoine Laurent Lavoisier and Armand Séguin, "Premier mémoire sur la respiration des animaux," 699.

31. Lavoisier and Séguin, "Premier mémoire sur la respiration des animaux," 699

32. Lavoisier and Séguin, "Premier mémoire sur la respiration des animaux," 698.

33. Lavoisier and Séguin, "Premier mémoire sur la respiration des animaux," 691. Cited in Holmes, *Lavoisier and the Chemistry of Life*, 449.

34. Antoine Laurent Lavoisier, "Altérations qu'éprouve l'air respiré," 676–87.

35. Lavoisier and Séguin, "Premier mémoire sur la respiration des animaux," 696.

36. Lavoisier and Séguin, "Premier mémoire sur la respiration des animaux," 697.

37. Holmes, *Lavoisier and the Chemistry of Life*, 455.

38. Lavoisier and Séguin, "Premier mémoire sur la respiration des animaux," 697.

39. Lavoisier and Séguin, "Premier mémoire sur la respiration des animaux," 696.

40. Lavoisier and Séguin, "Premier mémoire sur la respiration des animaux," 703.

41. Charles Coulston Gillispie, *Science and Polity in France*, 88–91.

42. Antoine Laurent Lavoisier, "Résultats extraits d'un ouvrage intitulé De la richesse territoriale du royaume de France," 407.

43. Norton Wise, "Work and Waste," 268; Bensaude-Vincent, "The Balance," 228.

44. Pierre-Simon Laplace, *Exposition du système du monde*, 76.

45. Jean-Claude Perrot, *Une histoire intellectuelle de l'économie politique, XVIIe–XVIIIe siècle*, 181.

46. Wise, "Mediations," 232.

47. Lavoisier and Séguin, "Premier mémoire sur la respiration des animaux," 700.

48. Lavoisier and Séguin, "Premier mémoire sur la respiration des animaux," 698.

49. Bourguinat, *Les grains du désordre*, 311.

50. Karl Marx, "Critique of the Gotha Program."

51. Lavoisier and Séguin, "Premier mémoire sur la respiration des animaux," 700.

52. Frederick L. Holmes, *Claude Bernard and Animal Chemistry*, 70.

53. Alan J. Rocke, "Organic Analysis in Comparative Perspective," 273–310.

54. McPhee, *The Politics of Rural Life*, 70.

55. Emile Zola's *Germinal*, for example, records the social disintegration of striking miners. "Such nastiness had never been seen before. Hunger exacerbated grudges; one felt the need to strike out, and a debate between two gossiping women would turn into a murder between two men." Emile Zola, *Germinal*, 1468.

56. Steven L. Kaplan, "The Famine Plot Persuasion in Eighteenth-Century France," 1–79.

57. Hubert Bonin, "Employment and the Revolution of 1848 in France."

58. Bourguinat, *Les grains du désordre*, 162-63.

59. Marcel Chaigneau, *Jean-Baptiste Dumas*, 296–97.

60. Ibid. 298.

61. Alan J. Rocke, *Nationalizing Science*, 113.

62. John Hubbel Weiss, *The Making of Technological Man*, 377. Dumas served for thirty-six years as president of the Society for the Encouragement of the National Industry. Chaigneau, *Jean-Baptiste Dumas*, 186.

63. Rocke, *Nationalizing Science*, 116.

64. Dumas's father-in-law was Alexandre Brogniart, himself an influential naturalist.

65. Jean Baptiste Boussingault, *Mémoires de J.-B. Boussingault*, 4:196. Cited in McCosh, *Boussingault*, 72.

66. McCosh, *Boussingault*, 73.

67. McCosh, *Boussingault*, 96.

68. Jean Baptiste Boussingault, *Economie rurale considérée dans ses rapports avec la chimie, la physique et la météorologie*, 298.

69. Pinkney, *Decisive Years in France, 1840–1847*, 63. This claim appears overly final, given the large-scale consumer movements during the First and Second World Wars.

70. Philippe Vigier, *1848, les français et la république*, 35–54.

71. "Troubles à Paris pour le pain," 202.

72. Bourguinat, *Les grains du désordre*, 444.

73. "Subsistances du peuple," 202.

74. "Proclamation concernant la garantie et l'organisation du travail du 25-29 février 1848.".

75. Jean Baptiste Dumas, *Traité de chimie appliquée aux arts*, 8:423.

76. Jean Baptiste Dumas and Jean Baptiste Boussingault, *Essai de statique chimique des êtres organisés*, 88.

77. Dumas, *Traité de chimie appliquée aux arts*, 8:457.

78. Dumas and Boussingault, *Essai de statique chimique des êtres organisés*, 88.

79. Anselme Payen, *Précis théorique et pratique des substances alimentaires et des moyens de les améliorer, de les conserver et d'en reconnaître les altérations*, 483.

80. Dumas, *Traité de chimie appliquée aux arts*, 8:458.

81. Félix Leblanc, "Recherches sur la composition de l'air confiné (Mémoire lu à l'Académie des Sciences, le 6 juin 1842)," 231; Jean Baptiste Boussingault, "Recherches sur la quantité d'acide carbonique contenue dans l'air de la ville de Paris," 458.

82. Boussingault, *Economie rurale considérée dans ses rapports avec la chimie, la physique et la météorologie*, 389.

83. The term "ration of air" is from Leblanc, "Recherches sur la composition de l'air confiné (Mémoire lu à l'Académie des Sciences, le 6 juin 1842)," 236.

84. Louis René Lecanu, "Nouvelles recherches sur l'urine humaine," 676–96.

85. Dumas, *Traité de chimie appliquée aux arts*, 8:423; Boussingault, *Economie rurale considérée dans ses rapports avec la chimie, la physique et la météorologie*, 399.

86. Jean Baptiste Boussingault, "Analyses comparées de l'aliment consommé et des excréments rendus par une tourterelle, entreprises pour rechercher s'il y a exhalation d'azote pendant la respiration des granivores," 434.

87. Jean Baptiste Dumas, *Eloge historique de Henri-Victor Regnault*, 20; Kenneth Carpenter, *Protein and Energy*, 70.

88. Anselme Payen, "De l'alimentation publique," 899.

89. Boussingault, *Economie rurale considérée dans ses rapports avec la chimie, la physique et la météorologie*, 293.

90. Boussingault, *Economie rurale considérée dans ses rapports avec la chimie, la physique et la météorologie*, 255.

91. Boussingault, *Economie rurale considérée dans ses rapports avec la chimie, la physique et la météorologie*, 293.

92. Dumas, *Traité de chimie appliquée aux arts*, 8:423.

93. For example, Conseil de santé des armées, I*nstruction à l'effet de guider les troupes dans la composition de leur régime alimentaire (5 mars 1850)*.

94. Dumas, *Traité de chimie appliquée aux arts*, 8:462–63.

95. Dumas, *Traité de chimie appliquée aux arts*, 8:423.

96. Dumas and Boussingault, *Essai de statique chimique des êtres organisés*, 7.

97. Dumas and Boussingault, *Essai de statique chimique des êtres organisés,* 7. See Argemí i d'Abadal, "Agriculture, Agronomy, and Political Economy," 459.

98. Dumas and Boussingault, *Essai de statique chimique des êtres organisés*, 7. See Simmons, "Waste Not, Want Not," 73–98.

99. E. C. Spary, "The Matter of Nourishment," in *Feeding France: New Sciences of Food, 1760-1815*, 97.

100. Holmes, *Claude Bernard and Animal Chemistry*, 20.

101. Holmes, *Claude Bernard and Animal Chemistry*, 34–103.

102. Persoz, "Expériences sur l'engrais des oies," 418.

103. Jean Baptiste Dumas and Henri Milne Edwards, "Sur la composition de la cire des abeilles," 407.

104. Dumas, *Traité de chimie appliquée aux arts*, 8:464.

105. Dumas, *Traité de chimie appliquée aux arts*, 8:418.

106. François Magendie, "Rapport fait à l'Académie des Sciences au nom de la Commission dite de la gélatine," 281.

107. Holmes, *Claude Bernard and Animal Chemistry*, 245.

108. Dumas, *Traité de chimie appliquée aux arts*, 8:424.

109. Dumas, *Traité de chimie appliquée aux arts*, 8:426.

110. Dumas, *Traité de chimie appliquée aux arts*, 8:424.

CHAPTER THREE

1. Jean Baptiste Dumas and Jean Baptiste Boussingault, *Essai de statique chimique des êtres organisés.*

2. Ann F. La Berge, *Mission and Method*, 2.

3. Andrew R. Aisenberg, *Contagion*, 24.

4. William Coleman, *Death Is a Social Disease*, 178.

5. Joshua Cole, *The Power of Large Numbers*, 17.

6. Gérard Jorland, *Une société à soigner*, 98, citing Louis-René Villermé, "Rapport sur une série de tableaux relatifs au mouvement de la population dans douze arrondissements municipaux de la ville de Paris," *Archives générales de médecine* X, 1826, 228.

7. Jorland, *Une société à soigner*, 102.

8. Coleman, *Death Is a Social Disease*, 178.

9. Louis-René Villermé, *Tableau de l'état physique et moral des ouvriers employés dans les manufactures de coton, de laine et de soie*, 16.

10. Villermé, *Tableau de l'état physique et moral des ouvriers employés dans les manufactures de coton, de laine et de soie*, 24.

11. Jorland, *Une société à soigner*, 110.

12. Aisenberg, *Contagion*, 44, 55.

13. Jorland, *Une société à soigner*, 89.

14. Jorland, *Une société à soigner*, 83.

15. Robert Castel, *Les métamorphoses de la question sociale*, 32.

16. Giovanna Procacci, *Gouverner la misère*, 16.

17. Pierre Honoré Bérard, "Rapport sur le régime alimentaire des lycées de Paris," 336.

18. Bérard, "Rapport sur le régime alimentaire des lycées de Paris, 336.

19. Pierre Honoré Bérard, *Cours de physiologie, fait à la Faculté de médecine de Paris*, 802.

20. Jorland, *Une société à soigner*, 89.

21. Bérard, *Cours de physiologie, fait à la Faculté de médecine de Paris*, 580.

22. Bérard, *Cours de physiologie, fait à la Faculté de médecine de Paris*, 584.

23. François Magendie, "Rapport fait à l'Académie des Sciences au nom de la Commission dite de la gélatine," 239.

24. Frederick L. Holmes, *Claude Bernard and Animal Chemistry*, 9.

25. Bernadette Bensaude-Vincent, "The Balance," 223.

26. Stephen Rössner, "The Wine Merchant as Weight Watcher," 577–78. Lucia Dacome, "Living with the Chair," 489.

27. Dacome, "Living with the Chair," 475.

28. E. C. Spary, *Feeding France*, 103.

29. J. B. Lawes, "Animal Chemistry," 459–544.

30. Ignatz Johann Heyss, *Der nach dem Lebengewicht des Rindes und selbst nach dessen Quadrat-Maaße normalmäßig zu bestimmende Milchertrag, oder das einfachste und sicherste Mittel zur höchsten Veredlung und Vervollkommung der Rindviehzucht und höchst möglichen Futterausnützung*, 50.

31. John Hubbel Weiss, *The Making of Technological Man*, 51–52.

32. Katherine Lynch, *Family, Class, and Ideology in Early Industrial France*, 16; Pamela Pilbeam, *The 1830 Revolution in France*, 95.

33. Pierre Rosanvallon, *Le moment Guizot*, 47.

34. Maurice Agulhon, *1848, ou l'apprentissage de la République*, 10.

35. Dominique Kalifa, "Conclusion," 303.

36. Ken Alder, "A Revolution to Measure," 39–71.

37. Bensaude-Vincent, "The Balance," 222.

38. Henri-Jean-Baptiste Davenne, *Mémoire au Conseil de surveillance sur la revision du régime alimentaire*, 32, was concerned with hospital meals.

39. Administration générale de l'assistance publique à Paris, *Régime alimentaire*, 8.

40. Bérard, "Rapport sur le régime alimentaire des lycées de Paris," 337.

41. Decrees of September 1, 1853, and February 13, 1854. Reprinted in A. Charles, *Législation des établissements publics d'instruction secondaire*, 252.

42. Bérard, "Rapport sur le régime alimentaire des lycées de Paris," 338.

43. Maxime Vernois, *Etat hygiénique des lycées de l'empire en 1867, extrait du rapport présenté à S. Exc. le Ministre de l'Instruction Publique*, 48.

44. Report of the Special Commission appointed by the prefect of the Seine to examine projects for the heating and ventilation of the Nouvelle Force prison. Cited in Philippe Grouvelle, *Chauffage et ventilation de la Nouvelle Force à Paris*, 24–28. See also Eugène Péclet, *Nouveaux documents relatifs au chauffage et à la ventilation des établissements publics*, 1–5.

45. Leblanc was one of several people endowed with "delicate noses" who served as subjects for this experiment, though it appears that only he endured the full ten-hour run.

46. In addition to Dumas and Boussingault, commissioners included the esteemed chemist Louis-Joseph Gay-Lussac, Claude Pouillet (physicist, meteorologist, telegraph enthusiast, and professor at the Conservatoire National des Arts et Métiers), two politicians, two members of the city's architectural counsel, and the prison's architects, Emile Gilbert and Jean François Lacointe. Grouvelle, *Chauffage et ventilation de la Nouvelle Force à Paris*, 36.

47. Grouvelle, *Chauffage et ventilation de la Nouvelle Force à Paris*, 28. The commission suggested that one raise this limit to 12m3 in summertime.

48. Jean Baptiste Dumas and Jean Baptiste Boussingault, *Essai de statique chimique des êtres organisés*, 119.

49. Jean Baptiste Boussingault, "Recherches expérimentales sur le développement de la graisse pendant l'alimentation des animaux," 423, and "Analyses comparées des alimens consommés et des produits rendus par une vache laitière," 118.

50. Boussingault, "Analyses comparées de l'aliment consommé et des excréments rendus par une tourterelle," 441.

51. Over the years, Péclet shifted from six to seven–eleven, then to twelve–fifteen cubic meters per hour. His popular works on heating and ventilation were widely read. Péclet, *Instruction sur l'assainissement des écoles primaires et des salles d'asile*, 16, and "Résumé des conditions du chauffage et de la ventilation des lieux habités," in *Traité de la chaleur considérée dans ses applications*, 606.

52. Maxime Vernois, *Etat hygiénique des lycées de l'empire en 1867*, 100.

53. Jean Baptiste Monfalcon and Augustin-Pierre Isidore de Polinière, *Traité de la salubrité dans les grandes villes suivi de l'hygiène de Lyon*, 64.

54. Michelle Perrot, "Les ouvriers, l'habitation et la ville au XIXe siècle," in *La question du logement et le mouvement ouvrier français*, 23.

55. Charles Jean Melchior de Vögué, speech on July 14, 1848. Cited in Emile de Girardin, *L'abolition de la misère par l'élevation des salaires*, 41.

56. Conseil de salubrité, "Instruction concernant les moyens d'assurer la salubrité des habitations," 612.

57. Conseil de salubrité, "Instruction concernant les moyens d'assurer la salubrité des habitations," 614. The prefect followed this wording exactly in his ruling. See Préfecture de la Seine, "Ordonnance concernant la salubrité des habitations," 608.

58. Péclet, *Nouveaux documents relatifs au chauffage et à la ventilation des établissements publics*, 17.

59. Commission des logements insalubres, Département de la Seine, Ville de Paris, "Rapport général des travaux de la Commission pendant l'année 1851," in *Rapports généraux sur les travaux de la Commission pendant les années 1851 à 1869*, 6, 8. The Consulting Committee on Public Hygiene, for its part, called for a rule restricting the number of inhabitants per room, but to no avail. "Projet d'instruction pour l'execution de la loi relative à l'assainissement des logements insalubres," 8 July 1850. Archives Nationales (AN) F8 210.

60. Ann-Louise Shapiro, *Housing the Poor of Paris, 1850–1902*, 18.

61. This debate was stimulated by Alexis de Tocqueville's investigative report on American carceral innovations. Christopher Mead, "Gilbert, Emile(-Narcisse-Jacques)," Grove Dictionary of Art Online," http://www.groveart.com.

62. "Project de loi sur la réforme des prisons." *Moniteur universel*, May 10, 1840, 1001. Cited in Charles Berriat-Saint-Prix, *Mazas: Etude sur l'emprisonnement individuel*, 7.

63. Maison centrale de Force et de Correction, *Analyse des réponses des directeurs une circulaire ministérielle du 10 mars 1834 sur l'effet des régimes de ces maisons*, 36.

64. Ministère de l'Intérieur, *Projet de loi sur les prisons*, ii.

65. The Conciergerie contained a few individual holding cells, and the La Roquette prison for youth had experimented with solitary confinement. Mazas, however, was the first large-scale

project of its kind in France. Benjamin Appert, *Bagnes, prisons et criminels* 1:155. Mead, "Gilbert, Emile(-Narcisse-Jacques)."

66. Berriat-Saint-Prix, *Mazas: Etude sur l'emprisonnement individuel*, 10.

67. Berriat-Saint-Prix, *Mazas: Etude sur l'emprisonnement individuel*, 17.

68. "Le ministère de l'intérieur," *Le Siècle*, 1.

69. In June 1850, the prefect appointed a second Mazas commission, which gave reports on the "moral" and "physical" conditions at the prison. Péclet, *Nouveaux documents relatifs au chauffage et à la ventilation des établissements publics*, 18–21.

70. Eugène Péclet, "Chauffage et ventilation," in *Traité de la chaleur considérée dans ses applications aux arts*, 226. This resolution did not stop the debate over solitary imprisonment. Still into the 1890s, arguments pitted the mental and physical anguish of isolated prisoners against the salutary effects of solitary contemplation and privacy. Critics included the former prison doctor at Mazas and one of its better-known residents, Jules Vallès. See Prosper de Pietra Santa, *Mazas: Etudes sur l'emprisonnement cellulaire*, 31.

71. Ambroise Tardieu, "Pénitentiaire (système)," 244.

72. A. Hurel, "Du régime alimentaire dans les maisons centrales," 348.

73. Hurel, "Du régime alimentaire dans les maisons centrales," 346.

74. Marc d'Espine, "Notice sur les variations du poids des prisonniers soumis au régime pénitentiaire," 71–85.

75. Tardieu, "Pénitentiaire (système)," 244.

76. "Instruction sur la répartition du produit du travail des condamnés, 28 mars 1844." Cited in Merry Delabost, *Régime pénitentiaire; hygiène alimentaire; l'alimentation des détenus au point de vue hygiénique et pénitentiaire*, 5.

77. "Circulaire sur les amélioration que pourrait exiger le régime alimentaire dans les maisons centrales, par suite de l'Ordonnance du 27 septembre 1843." Cited in Merry Delabost, *Régime pénitentiaire; hygiène alimentaire; l'alimentation des détenus au point de vue hygiénique et pénitentiaire*, 6.

78. Ambroise Tardieu, "Subsistances,"202–15.

79. Alfred Becquerel, *Traité élémentaire d'hygiène privée et publique*, 425–30.

80. Payen is best known for his improvements in chemical industrial processes such as the production of borax, for having identified vegetable cellulose, and for having first isolated and artificially produced the enzyme that broke starch down into sugar. See F. W. J. McCosh, *Boussingault*, 280.

81. Anselme Payen, *Précis théorique et pratique des substances alimentaires et des moyens de les améliorer, de les conserver et d'en reconnaître les altérations*, vii.

82. Payen, *Précis théorique et pratique des substances alimentaires et des moyens de les améliorer, de les conserver et d'en reconnaître les altérations*, 14.

83. Payen, *Précis théorique et pratique des substances alimentaires et des moyens de les améliorer, de les conserver et d'en reconnaître les altérations*, 502.

84. Jean Boulaine and Jean-Paul Legros, "Gasparin," 60.

85. Adrien de Gasparin, *Cours d'agriculture* 3:58.

86. Prison doctor Merry Delabost cites Gasparin in referring to the two types of ration. Delabost, *Régime pénitentiaire; hygiène alimentaire; l'alimentation des détenus au point de vue*

hygiénique et pénitentiaire, 17. Military doctor L. Kirn identifies Gasparin as the originator of the work and maintenance model. L. Kirn, "L'alimentation du soldat," 281. See also Jane O'Hara-May, "Measuring Man's Needs," 262.

87. Gasparin, *Cours d'agriculture*, 3:52.

88. Gasparin, *Cours d'agriculture*, 3:51.

89. François Vatin, *Le travail, économie et physique 1780–1830*, 10.

90. Charles Augustin Coulomb, "Résultat de plusieurs expériences destinées à déterminer la quantité d'action que les hommes peuvent fournir par leur travail journalier, suivant les différentes manières dont ils emploient leurs forces," 380–428.

91. William Reddy, *The Rise of Market Culture*, 141.

92. Vatin, *Le travail, économie et physique 1780–1830*, 104.

93. Gasparin, *Cours d'agriculture*, 48.

94. Gasparin, *Cours d'agriculture*, 3:54.

95. Gasparin, *Cours d'agriculture*, 3:54.

96. Jean Baptiste Dumas, *Traité de chimie appliquée aux arts*, 424.

97. Gasparin, *Cours d'agriculture*, 3:32.

98. Gasparin, *Cours d'agriculture*, 3:33.

99. Gasparin, *Cours d'agriculture*, 3:36.

100. Gasparin, *Cours d'agriculture*, 3:34.

101. Gasparin, *Cours d'agriculture*, 3:33.

102. Gasparin, *Cours d'agriculture*, 3:49–50.

103. Gasparin, *Cours d'agriculture*,3:56.

104. Gasparin, *Cours d'agriculture*, 3:50–51.

105. Gasparin, *Cours d'agriculture*, 3:34.

106. Gasparin, *Cours d'agriculture*, 3:54.

107. Gasparin, *Cours d'agriculture*, 3:58.

108. Gasparin, *Cours d'agriculture*, 3:34.

109. Charles Schindler, *L'alimentation du soldat en campagne*, 10.

110. Gasparin, *Cours d'agriculture*, 49.

111. Gasparin, *Cours d'agriculture*, 3:52.

112. Gasparin, *Cours d'agriculture*, 3:54.

113. Antoine Laurent Lavoisier, "Résultats extraits d'un ouvrage intitulé De la richesse territoriale du royaume de France," 413.

114. Gasparin, *Cours d'agriculture*, 3:55.

115. Gasparin, *Cours d'agriculture*, 3:63.

116. Gasparin, *Cours d'agriculture*, 3:57.

CHAPTER FOUR

1. Schacherer, "Budget des travailleurs," 6.

2. Robert Castel, *Les métamorphoses de la question sociale*, 238–42.

3. Eileen Janes Yeo, "Social Surveys in the Eighteenth and Nineteenth Centuries," 91.

4. Alain Desrosières, *The Politics of Large Numbers*, 33, 42.

5. Peter Wagner, "'An Entirely New Object of Consciousness, of Volition, of Thought': The Coming into Being and (Almost) Passing Away of 'Society' as a Scientific Object," in *Biographies of Scientific Objects*, 133.

6. Jean-Claude Perrot, *Une histoire intellectuelle de l'économie politique, XVIIe–XVIIIe siècle*, 185.

7. One of the first systematic manuals described a voyage within France itself. See Jean-Luc Chappey, *La Société des Observateurs de l'Homme (1799–1804)*, 263.

8. Wagner, "An Entirely New Object of Consciousness, of Volition, of Thought," 135.

9. The term makes reference to Elizabeth Williams's description of a French "anthropological medicine." Elizabeth A. Williams, *The Physical and the Moral*, 1.

10. George Stocking, "French Anthropology in 1800," 134–50.

11. Castel, *Les métamorphoses de la question sociale*, 248.

12. Paul Rabinow alludes to the co-emergence of anthropology and welfare in Paul Rabinow, *French Modern*, 24.

13. Frédéric Le Play, *Instruction sur la méthode d'observation dite des monographies de familles, propre à l'ouvrage intitulé "Les ouvriers européens,"* 13.

14. Georges Cuvier, *Le règne animal distribué d'après son organisation*, 6.

15. Dorinda Outram, "Uncertain Legislator," 331.

16. Georges Cuvier, *Leçons d'anatomie comparée*, 57.

17. Cuvier, *Leçons d'anatomie comparée*, 53.

18. Cuvier, *Leçons d'anatomie comparée*, 53.

19. Pierre-Jean-Georges Cabanis, *Rapports du physique et du moral de l'homme*, 224.

20. Williams, *The Physical and the Moral*, 93.

21. Sean Quinlan, "Physical and Moral Regeneration after the Terror," 150–51.

22. Williams, *The Physical and the Moral*, 100.

23. Quinlan, "Physical and Moral Regeneration after the Terror," 164.

24. Jean Baptiste Lamarck, *Philosophie zoologique*, 8.

25. Lamarck, *Philosophie zoologique*, 325.

26. Lamarck, *Philosophie zoologique*, 351.

27. Lamarck, *Philosophie zoologique*, 357.

28. Christian Bange, Pietro Corsi, and Pascal Duris, "Les médecins auditeurs du cours professé par Lamarck au Muséum (1795–1823)," 399, 404.

29. M. Jomard, *Discours sur la vie et les travaux du Baron de Gérando*, 17–19.

30. Dorinda Outram, *Georges Cuvier: Vocation, Science, and Authority in Post-Revolutionary France*, 55.

31. Chappey, *La Société des Observateurs de l'Homme (1799–1804)*, 135.

32. George Weisz, *The Medical Mandarins*, 17.

33. Jomard, *Discours sur la vie et les travaux du Baron de Gérando*, 25.

34. Quinlan, "Physical and Moral Regeneration after the Terror," 164.

35. Stocking, "French Anthropology in 1800," 134–50.

36. Joseph Degerando, "Considérations dans les diverses méthodes à suivre dans l'observation des peuples sauvages," 154.

37. Degerando, "Considérations dans les diverses méthodes à suivre dans l'observation des peuples sauvages," 154.

38. Degerando, "Considérations dans les diverses méthodes à suivre dans l'observation des peuples sauvages," 155.

39. Degerando, "Considérations dans les diverses méthodes à suivre dans l'observation des peuples sauvages," 155.

40. Cuvier, *Le règne animal distribué d'après son organisation*, 6.

41. Degerando, "Considérations dans les diverses méthodes à suivre dans l'observation des peuples sauvages," 155. Emphasis mine.

42. Degerando, "Considérations dans les diverses méthodes à suivre dans l'observation des peuples sauvages," 155.

43. Degerando, "Considérations dans les diverses méthodes à suivre dans l'observation des peuples sauvages," 165.

44. Degerando, "Considérations dans les diverses méthodes à suivre dans l'observation des peuples sauvages," 171.

45. Degerando, "Considérations dans les diverses méthodes à suivre dans l'observation des peuples sauvages," 155.

46. Chappey, *La Société des Observateurs de l'Homme (1799–1804)*, 267.

47. Chappey, *La Société des Observateurs de l'Homme (1799–1804)*, 266.

48. Chappey, *La Société des Observateurs de l'Homme (1799–1804)*, 268.

49. Joseph Marie de Gérando, *De l'indigence, des causes de l'indigence, des devoirs imposés à la bienfaisance publique*, 24.

50. Gérando instructs the observer "to penetrate" his subject in both sets of documents: Degerando, "Considérations dans les diverses méthodes à suivre dans l'observation des peuples sauvages," 155; Joseph Marie de Gérando, *Le visiteur du pauvre*, 34.

51. Gérando, *De l'indigence, des causes de l'indigence, des devoirs imposés à la bienfaisance publique*, 24–25.

52. Gérando, *Le visiteur du pauvre*, 34.

53. Gérando, *De l'indigence, des causes de l'indigence, des devoirs imposés à la bienfaisance publique*, 22.

54. Gérando, *Le visiteur du pauvre*, 46.

55. Gérando, *Le visiteur du pauvre*, 35. Emphasis mine.

56. Desrosières, *The Politics of Large Numbers*, 213.

57. Gérando, *Le visiteur du pauvre*, 123.

58. Gérando, *De l'indigence, des causes de l'indigence, des devoirs imposés à la bienfaisance publique*, 26.

59. Gérando, *De l'indigence, des causes de l'indigence, des devoirs imposés à la bienfaisance publique*, 34. A survey of workers' councils (Conseils de Prudhommes) by the Prefecture of the Seine concerning wages from 1830 to 1847 gives a rough sense of where Gérando's figures stood with respect to actual pay rates in urban Paris. Reported wages of unskilled manual laborers (*hommes de peine*), among the least paid workers, ranged from 2.25 to 2.75 francs per day. Jacques Rougerie, "Remarques sur l'histoire des salaires à Paris au XIXe siècle," 75. Rougerie cites the Archives Départementales de la Seine, DM12(23), "Conseils de Prudhommes: Salaires de 1846."

60. Gérando, *De l'indigence, des causes de l'indigence, des devoirs imposés à la bienfaisance publique*, 26.

61. Emile Cheysson, *Frédéric Le Play, sa doctrine, sa méthode, son école*, 10.

62. Cheysson, *Frédéric Le Play, sa doctrine, sa méthode, son école*, 9.

63. Theodore M. Porter, "Reforming Vision: the Engineer Le Play Learns to Observe Society Sagely," in *Histories of Scientific Observation*, 283–85.

64. Frédéric Le Play, *L'organisation du travail selon la coutume des ateliers et la loi du Décalogue*, 144.

65. Frédéric Le Play, "Statistique," 275–77.

66. Le Play, *Instruction sur la méthode d'observation dite des monographies de familles, propre à l'ouvrage intitulé "Les ouvriers européens,"* 14.

67. Le Play, "Statistique," 275.

68. Le Play, *Instruction sur la méthode d'observation dite des monographies de familles, propre à l'ouvrage intitulé "Les ouvriers européens,"* 14.

69. Le Play, *Instruction sur la méthode d'observation dite des monographies de familles, propre à l'ouvrage intitulé "Les ouvriers européens,"* 30.

70. Katherine Lynch, *Family, Class, and Ideology in Early Industrial France*, 45.

71. Sharon Marcus, *Apartment Stories*, 153–56; Emile Cheysson, *Société anonyme de Passy-Auteuil pour les Habitations Ouvrières*, 16.

72. Lynch, *Family, Class, and Ideology in Early Industrial France*, 45.

73. Wally Seccombe, *Weathering the Storm*, 8; Giovanna Procacci, *Gouverner la misère*, 164.

74. Le Play, *Instruction sur la méthode d'observation dite des monographies de familles, propre à l'ouvrage intitulé "Les ouvriers européens,"* 29.

75. One complete list of a Le Playist typology can be found in Congrès international de statistique, "Résolutions adoptées à Bruxelles . . . ," 14–16.

76. Frédéric Le Play, "The Study of Working Class Families," in *Frédéric Le Play On Family, Work, and Social Change*, 172. Translated by Catherine Bodard Silver.

77. Frédéric Le Play, *Les ouvriers européens*, 81.

78. Le Play, *Les ouvriers européens*, 580–81.

79. Le Play, *Les ouvriers européens*, 360.

80. Philippe Le Gall, *A History of Econometrics in France*, 5.

81. See chap. 6 below.

82. Desrosières, *The Politics of Large Numbers*, 214.

83. Emile Cheysson and Alfred Toqué, *Les budgets comparés des cent monographies de familles*, 29.

84. Emile Cheysson, "La loi de l'hérédité," 3.

85. Schacherer, "Budget des travailleurs," 5.

86. Schacherer, "Budget des travailleurs," 5.

87. Schacherer, "Budget des travailleurs," 3.

88. Schacherer, "Budget des travailleurs," 5.

89. Schacherer, "Budget des travailleurs," 6.

90. Victoria E. Thompson, *The Virtuous Marketplace*, 59–65.

91. Schacherer, "Budget des travailleurs," 10.

92. Eugène Buret, *La misère des clases laborieuses en Angleterre et en France*, 111.

93. Jean Charles Léonard de Sismondi, *Nouveaux principes d'économie politique*, 79.

94. Procacci, *Gouverner la misère*, 145.

95. Margaret Schabas, *The Natural Origins of Economics*, 2.

96. Sismondi, *Nouveaux principes d'économie politique*, 345.

97. Sismondi, *Nouveaux principes d'économie politique*, 9.

98. Charles Renouvier, *Manuel républicain de l'homme et du citoyen*, 105. The November Constitution, passed in the wake of a conservative electoral victory, retained a right to existence only for "citizens in need" and proposed a plan for poor aid in lieu of guaranteed employment. Castel, *Les métamorphoses de la question sociale*, 274.

99. Louis Blanc, *La révolution de fevrier au Luxembourg*, 70–71.

100. Louis Blanc, *Organisation du travail*, 3.

101. R. Gossez, *Les ouvriers de Paris*, 66.

102. Gossez, *Les ouvriers de Paris*, 78.

103. See Sewell's persuasive account of this unintentional development in William H. Sewell, Jr., *Work and Revolution in France*, 258.

104. Gossez, *Les ouvriers de Paris*, 66.

105. Assemblée Constituante, Correspondance Générale: Prud'hommes, 7 March 1849, délégués des ouvriers couvreurs, 139, 140 (Archives Nationales C 2277). Cited in Gossez, *Les ouvriers de Paris*, 76.

106. Gossez cites five similar petitions. Gossez, *Les ouvriers de Paris*, 77.

107. *Salut du Peuple*, December 1849. Cited in Gossez, *Les ouvriers de Paris*, 77.

CHAPTER FIVE

1. G. Dillon-Kavanagh, "La ration," *La Patrie en Danger*, October 6, 1870.

2. Auguste Blanqui, "Le rationnement," *La Patrie en Danger*, December 8, 1870.

3. G. de Molinari, *Journal des débats*, October 11, 1870.

4. F. Bresson, "La question de la viande," *Le Figaro,* October 28, 1870.

5. Molinari. *Journal des débats*, October 21, 1870.

6. Roger V. Gould, *Insurgent Identities*, 5, 137–41, 156.

7. Jules Favre, vice-president of the Government of National Defense, reflected this sentiment in a speech to the Council of War on December 31, 1870: "The city of Paris wants to be defended à outrance." Cited and translated by Robert Tombs, "The Wars against Paris," 544.

8. The Prussians' clear intentions on this score were revealed by Bismarck's refusal to allow provisions to enter Paris as part of negotiations for a temporary armistice in November 1870. See Tombs, "The Wars against Paris," 549–50.

9. Paul Vesson, letter dated September 11, 1870, in Paul Bordarier, "Le siège de Paris d' après un recueil de lettres-ballons," *La Revue Hommes et Mondes* 10, no. 39 (1949): 262.

10. William Serman, "French Mobilization in 1870," in Förster and Nagler, On the Road to Total War," 289.

11. De Molinari, *Journal des débats*, October 21, 1870. De Molinari here attributes this image of Paris to Auguste Blanqui.

12. "Jeudi soir, 15 septembre. Club de la Patrie en Danger, Salle Favié, à Belleville," *Journal des débats*, September 20, 1870.

13. "Club de La Patrie en Danger, Présidence du Citoyen Blanqui: Séance du 12 septembre 1870," *La Patrie en Danger*, September 14, 1870.

14. Jean Tapié and Ménier Tapié, "Les mesures de salut public," October 1870, 1.

15. Tapié and Tapié, *Les mesures de salut public*, 1

16. Tapié and Tapié, *Les mesures de salut public*, 1

17. "Club de la Patrie en Danger, rue d'Arras," *Journal des débats,* September 26, 1870.

18. A. L.,"Les bouches inutiles," 1870.

19. Edme Bourgoin, *Hygiène publique*, 18.

20. Ernest Denormandie, *Notes et Souvenirs par M. Denormandie, Senateur. Les Journées de Juin 1848 - Le Siège de Paris - La Commune - L'Assemblée Nationale - Quelques Réflexions* (Paris: Société Anonyme des Publications Périodiques, 1896).

21. As compared to 1869. Bertrand Taithe, *Defeated Flesh: Medicine, Welfare and Warfare in the Making of Modern France* (Lanham: Rowman & Littlefield, 1999), 122.

22. Adolphe Morillon, *L'approvisionnement de Paris,* 120.

23. Taithe, *Defeated Flesh,* 64. The Government of National Defense responded by firing the director of the Assistance Publique, Armand Husson.

24. Dillon-Kavanagh, "La ration."

25. For a wonderful analysis of social life during the siege, see Hollis Clayson, *Paris in Despair: Art and Everyday Life under Siege (1870–71)* (Chicago: University of Chicago Press, 2002). Clayson's work was invaluable for this chapter.

26. Jules Favre et al., "Aux habitans de Paris," *Journal des débats,* December 13, 1870.

27. "On lit dans le Bulletin de la municipalité de Paris," *Journal des débats,* September 28, 1870.

28. P. David, "Nous aimerions savoir en vertu de quel mandat . . ." *Journal des débats,* October 4, 1870.

29. Bertrand Taithe, *Citizenship and Wars*, 45.

30. Rebecca Spang, " "And They Ate the Zoo": Relating Gastronomic Exoticism in the Siege of Paris," MLN 107, no. 4, French Issue (1992): 759.

31. Stéphane Rials, *Nouvelle histoire de Paris: de Trochu à Thiers, 1870–1873,* ed. Association pour la publication d'une histoire de Paris, Nouvelle histoire de Paris (Paris: Hachette, 1985), 201.

32. Jean-Jules Clamageran, *Souvenirs du siège de Paris*, 12.

33. Morillon, *L'approvisionnement de Paris,* 125.

34. Morillon, *L'approvisionnement de Paris,* 124.

35. Morillon, *L'approvisionnement de Paris*, 120; Clamageran, *Souvenirs du siège de Paris*, 17–18.

36. De Molinari, *Journal des débats,* October 21, 1870.

37. De Molinari, *Journal des débats,* October 21, 1870.

38. G. de Molinari, "L'alimentation de Paris pendant le siège," *Revue des Deux-Mondes*, 41st year - second period, no. 91 (1871): 121.

39. Clamageran, *Souvenirs du siège de Paris*, 14–17.

40. Juliette Lamber Adam, *Mes illusions et nos souffrances pendant le siège de Paris,* Sixth ed. (Paris: Alphose Lemerre, 1906).

41. Albert Goullé, "La question des vivres," *La Patrie en Danger*, September 14, 1870.

42. Steven L. Kaplan, *Le retour du bon pain.*

43. Since the 1850s, a union of bakers had agreed on a common price with the blessing of the government. Morillon, *L'approvisionnement de Paris*, 143.

44. Jules Ferry, in *Enqûete parlementaire sur les Actes du Gouvernement de la Défense nationale: déposition des témoins* (Versailles, 1873), 1:421. Cited in Michael Howard, *The Franco-Prussian War: The German Invasion of France, 1870–1871* (London: Rupert Hart-Davis, 1962), 326.

45. Morillon, *L'approvisionnement de Paris,* 151. The most drastic changes in bread occurred in December 1870 and January 1871, near the end of the siege.

46. Pictures of such souvenirs can be found in the Archives Nationales (AB XIX 4005).

47. "La question du pain," *Le Figaro,* December 18, 1870.

48. Henri D'Alméras, *La vie parisienne pendant le siège et sous la commune*, 4.

49. Ernest-Guillaume Cresson, *Cent jours de siége à la préfecture de police* (Paris: Plon, Nourrit et Cie, 1901), 114.

50. Le Pourvoyeur, "Le pain: derniers incidents," *Le Figaro*, December 15, 1870.

51. Favre et al., "Aux habitans de Paris."

52. Cresson, *Cent jours de siége à la préfecture de police,* 114.

53. Morillon, *L'approvisionnement de Paris*, 156–59.

54. Armand Husson recorded a daily average of 432 grams of bread per person in Paris in the years 1872–73. Husson, *Les Consommations de Paris*, 517–19.

55. Morillon, *L'approvisionnement de Paris,* 165.

56. Letter dated January 19, 1871. Bordarier, "Le siège de Paris d' après un recueil de lettres-ballons," 272.

57. P. David, "Hier soir: le club Favié . . ." *Journal des débats,* January 18, 1871.

58. David, "Hier soir."

CHAPTER SIX

1. Lambeau, Monographies municipales de la ville de Paris: conditions du travail dans les chantiers communaux (1896), 1052. Cited in Barthélemy Raynaud, *Vers le salaire minimum*, 58.

2. Auguste Bouge, "Les conditions du travail et le collectivisme," 246.

3. Jean Jaurès, *Etudes socialistes*, xxxv, xlix.

4. Jaurès, *Etudes socialistes*, xxxvii.

5. See the following three chapters; and see Dutton on the central role of the *salaire vital* in shaping the welfare state after 1914. Paul V. Dutton, *Origins of the French Welfare State*, 251.

6. "Vital, ale," in *Dictionnaire de L'Académie française*, 730.

7. Elizabeth A. Williams, *The Physical and the Moral*, 11.

8. William H. Schneider, *Quality and Quantity*, 85–92.

9. Jack D. Ellis, *The Physician-Legislators of France*, 2.

10. Gérard Jorland, *Une société à soigner*, 302–15.

11. Andrew R. Aisenberg, *Contagion*, 130.

12. Two exceptions were a law prohibiting child labor and another instituting hygienic inspections of working conditions in factories.

13. Jules Guesde, *La loi des salaires et ses conséquences, suivi d'une réponse à la "Réponse à M. Clemenceau,"* 8.

14. "Letter to M. Clemenceau, Deputy from Montmartre, editor of the journal La Justice" (1881), in Guesde, *La loi des salaires et ses conséquences*, 28.

15. Guesde, *La loi des salaires et ses conséquences*, 17.

16. Karl Marx, *Capital*, 627. Elsewhere Marx emphasized that the expansion of basic needs is essential for capitalism to develop historically.

17. Guesde, *La loi des salaires et ses conséquences*, 6.

18. Ferdinand Lassalle, *Zur Arbeiterfrage*, 5.

19. Lassalle, *Zur Arbeiterfrage*, 9.

20. John Bellamy Foster, *Marx's Ecology*, 155.

21. Louis-René Villermé, *Tableau de l'état physique et moral des ouvriers employés dans les manufactures de coton, de laine et de soie*, 7.

22. Guesde, *La loi des salaires et ses conséquences*, 12.

23. Kenneth Carpenter, *Protein and Energy: A Study of Changing Ideas in Nutrition*, 89–90; Corinna Treitel, "Max Rubner and the Biopolitics of Rational Nutrition," 1–25.

24. Armand Gautier, *L'alimentation et les régimes chez l'homme sain et chez les malades*, 4.

25. Dr. Segard, "Hygiène et solidarité," 2.

26. Jean-Charles Roux, *Le régime alimentaire des écoliers au point de vue pratique*, 5.

27. See, for example, Hermann Schmitt, *Die Steuerfreiheit des Existenzminimums*, 86.

28. Jaurès, *Etudes socialistes*, xxxviii, l.

29. Jules Guesde and Paul Lafargue, *Le programme du Parti ouvrier*, 59.

30. Guesde and Lafargue, *Le programme du Parti ouvrier,* 60.

31. *Sixième Congrès national des syndicats de France*, 82.

32. Ladislas Mysyrowicz, "Karl Marx, la Première Internationale et la statistique," 56. Citing speech by Eugène Dupont to the Geneva Congress, 1866.

33. Marilyn Boxer, "Protective Legislation and the Marginalization of Women Workers," 45–66.

34. James Guillaume, *L'Internationale*, 8.

35. Mysyrowicz, "Karl Marx, la Première Internationale et la statistique," 84.

36. Antoine Savoye, "La monographie sociologique," 35.

37. *Sixième Congrès national des syndicats de France*, 81.

38. Michelle Perrot, *Les ouvriers en grève*, 133, citing Archives of the Préfet de Police, BA 169/17.

39. Perrot, *Les ouvriers en grève,* 268, citing *L'Intransigeant*, December 24, 1881.

40. Perrot, *Les ouvriers en grève,* 133, citing *Le Citoyen et la Bataille*, March 19, 1883.

41. Jacques Rougerie, "Remarques sur l'histoire des salaires à Paris au XIXe siècle," 97.

42. Olivier Marchand and Claude Thélot, *Le travail en France, 1800–2000*, 38.

43. Jean-Charles Asselin, "La stagnation économique," 208.

44. Judith Coffin, "Consumption, Production and Gender," 111–41. Victoria De Grazia emphasizes the rarity of credit in interwar Europe. Victoria De Grazia, *Irresistible Empire*, 90.

45. Sabine Rudischhauser, "Salaire minimum et libre concurrence," 202.

46. Rudischhauser, "Salaire minimum et libre concurrence," 203.

47. Minimum wage clauses were voted by town councils in Limoges (1891), Toulon (1891), Roubaix (1891), Albi (1894), and Toulouse (1896). Raynaud, *Vers le salaire minimum*, 66.

48. Alain Desrosières, *The Politics of Large Numbers*, 12.

49. Cernesson, Rapport. Cited in Bouge, "Les conditions du travail et le collectivisme," 246.

50. Desrosières, *The Politics of Large Numbers*, 213.

51. Desrosières, *The Politics of Large Numbers*, 96.

52. Isidore Finance, speech at Congrès ouvrier, Marseille, October 1879. Cited in Michelle Perrot, "Note sur le positivisme ouvrier," 203.

53. Emile Laporte, Fabien Magnin, and Isidore Finance, *Le positivisme au Congrès ouvrier*, 39.

54. Desrosières, *The Politics of Large Numbers*, 153.

55. Robert Salais and Jean Luciani, "Matériaux pour la naissance d'une institution," 83–108.

56. Salais and Luciani, "Matériaux pour la naissance d'une institution," 83–108.

57. Desrosières, *The Politics of Large Numbers*, 213.

58. Desrosières, *The Politics of Large Numbers*, 157–58.

59. Christian Baudelot and Roger Establet, *Maurice Halbwachs*, 37.

60. Maurice Halbwachs, "Les besoins et les tendances dans l'économie sociale," 185.

61. Halbwachs, *L'évolution des besoins dans les classes ouvrières*, 8. On the development of Halbwachs's theory of need from 1913 to 1933, see the marvelous Judith Coffin, "A Standard of Living?," 6–26.

62. Victoria De Grazia has an excellent discussion of Halbwachs's confrontation with the ILO study of American standard of living in De Grazia, *Irresistible Empire*, 122–24.

63. Halbwachs, *L'évolution des besoins dans les classes ouvrières*, 107.

64. Halbwachs, *L'évolution des besoins dans les classes ouvrières*, 132.

65. Halbwachs, *L'évolution des besoins dans les classes ouvrières*, 106.

66. Sanford Elwitt, *The Third Republic Defended*, 58.

67. Conseil Supérieur du Travail, *Septième session (Décembre 1897)*, 298.

68. Académie des Sciences Morales et Politiques, "Emile Cheysson."

69. Robert Burton Elekund and Robert F. Hébert, *Secret Origins of Modern Microeconomics*, 420.

70. See chapter 4, "Family, Race, Type."

71. Schneider, *Quality and Quantity*, 107.

72. Jacques Donzelot, *The Policing of Families*, 175.

73. Emile Cheysson, *La question de la dépopulation en France*, 14.

74. Janet R. Horne, *A Social Laboratory for Modern France*, 231.

75. Allyre Chassevant, *Le régime alimentaire de guerre*, 4.

76. Emile Cheysson and Alfred Toqué, *Les budgets comparés des cent monographies de familles*, 30.

77. Cheysson and Toqué, *Les budgets comparés des cent monographies de familles*, 31.

78. Emile Cheysson, *Le salaire au point de vue statistique, économique et social*, 14.

79. Cheysson and Toqué, *Les budgets comparés des cent monographies de familles*, 31.

80. On Durkheim's sociology of consumption, see Rosalind Williams, *Dream Worlds*, 322–45.

81. Emile Durkheim, *De la division du travail social*, 265.

82. Emile Durkheim, *Le suicide, étude de sociologie*, 285.

83. Durkheim, *Le suicide, étude de sociologie*, 285.

84. Durkheim, *De la division du travail social*, 280.

85. Durkheim, *Le suicide, étude de sociologie*, 273.

86. Durkheim, *Le suicide, étude de sociologie*, 275.

87. Durkheim, *Le suicide, étude de sociologie*, 276.

88. Durkheim, *Le suicide, étude de sociologie*, 283.

89. Durkheim, *Le suicide, étude de sociologie*, 274.

90. See "Languages of Luxury in Eighteenth-Century France," in John Shovlin, "The Cultural Politics of Luxury in Eighteenth-Century France," 577–606.

91. Bouge, Les conditions du travail et le collectivisme, 289.

92. Asselin, "La stagnation économique," 203.

93. Asselin, "La stagnation économique," 203.

94. Herman Lebovics, *The Alliance of Iron and Wheat in the Third French Republic, 1860–1914*, 135.

95. Economic historians estimate that 15 percent of consumer spending was redirected to domestic products by tariffs. Michel Lescure, "Le manque de ressources," 259.

96. Emile Cacheux, *L'économiste pratique*, 354.

97. Jules Moureau, *Le salaire et des associations coopératives*, 208.

98. Charles Lavollée, *Exposition universelle internationale de 1889 à Paris. Economie sociale Section I. Rapport de M. Charles Lavollée*, 11.

99. Henri Baudrillart, *Philosophie de l'économie politique*, 472.

100. Henri Baudrillart, "Le luxe et les moralistes," 362.

101. Baudrillart, *Philosophie de l'économie politique*, 473. Emphasis mine.

102. Merry Delabost, *Régime pénitentiaire*, 40.

103. Onésime Perrine, *La vie saine et à bon marché*, 6.

104. Tyler Stovall, "From Red Belt to Black Belt, 351–70.

105. L. Pascault, "Critique des méthodes employées pour établir la norme alimentaire," 206.

106. L. Perrier, "Rapport sur l'alimentation dans l'armée," 482.

107. Pascault, "Critique des méthodes employées pour établir la norme alimentaire," 34.

108. L. Landouzy, "De l'irrationnel et de l'insuffisant dans l'alimentation des ouvriers et des employés parisiens," 34. Landouzy's participation in the eugenic movement is detailed in Schneider, *Quality and Quantity*, 85–86.

109. Landouzy, "De l'irrationnel et de l'insuffisant dans l'alimentation des ouvriers et des employés parisiens," 39. A similar sentiment appeared in G. Bardet, *La ration normale*, 70–72.

110. E. Maurel, *Traité de l'alimentation et de la nutrition à l'état normal et pathologique*, 4.

111. Edmond Villey, *La question des salaires ou la question sociale*, 63.

112. Villey, *La question des salaires ou la question sociale*, 223–24.

113. Leo XIII, "Rerum Novarum. On the Condition of the Working Classes."

114. Cited in Raynaud, *Vers le salaire minimum*, 16.

115. Raynaud, *Vers le salaire minimum*, 22. Citing Henri Lorin, "Déclaration de la semaine sociale de Marseille" (1908).

116. Peter J. Bernardi, *Maurice Blondel, Social Catholicism, and Action Française*, 22.

117. Monseigneur Turinaz, *Le salaire familial et le Cardinal Zigliara*, 8-17.

118. L. Jardin, "Le juste salaire," 12.

119. Charles Perin, "Le juste salaire d'après l'encyclique *Rerum novarum*," 416.

120. Jardin, "Le juste salaire," 12.

121. Joan Scott points to the ties that nineteenth-century economists drew between improvidence, poverty, and crime, particularly prostitution. Joan Wallach Scott, " 'L'ouvrière! Mot impie, sordide,' " 143.

122. Bouge, "Les conditions du travail et le collectivisme," 289.

123. Villey, *La question des salaires ou la question sociale*, 228. Essentially the same formula is repeated by Georges Picot, *Les moyens d'améliorer la condition de l'ouvrier*, 7.

124. Lisa Tiersten, *Marianne in the Market*, 186–87.

125. Cheysson, *Le salaire au point de vue statistique, économique et social*, 16.

126. Cheysson, *Le salaire au point de vue statistique, économique et social*, 17.

127. Cheysson, *Le salaire au point de vue statistique, économique et social*, 18.

128. Leon Harmel's factory in the Val-de-Bois is commonly cited as a predecessor for family allowances. For a breakdown of family allowances by date and company, see Jean Pinte, *Les allocations familiales*, 69.

129. A record of administrative family allowances is in William Oualid and Charles Picquenard, *La guerre et le travail*, 242. Dutton notes that family allowances had been distributed by the French army in times of war since the Revolution. See Dutton, *Origins of the French Welfare State*, 17.

130. Emile Romanet, *Le salaire familial*, 24.

131. Susan Pedersen, *Family, Dependence, and the Origins of the Welfare State*, 226.

132. Pedersen, *Family, Dependence, and the Origins of the Welfare State*, 413.

133. Georges Bonvoisin, *Allocations familiales et caisses de compensation: leur origine, leur raison d'être, leurs effets, leur fonctionnement dans la Région Parisienne*, 8.

134. Roger Picard, *Le salaire et ses compléments*, 59.

135. Pedersen, *Family, Dependence, and the Origins of the Welfare State*, 225.

136. Oualid and Picquenard, *La guerre et le travail*, 217.

137. Jonathan Manning, "Wages and Purchasing Power," 256.

138. Oualid and Picquenard, *La guerre et le travail*, 184–85; Dutton, *Origins of the French Welfare State*, 16–20.

139. John Barzman, "La 'crise de la vie chère' de l'été 1919 au Havre," 61–84.

140. Oualid and Picquenard, *La guerre et le travail*, 21.

141. "Budgets et salaires," 6.

142. This claim is based on a survey of articles on the subject in the workers' newspaper *L'Information ouvrière et sociale*, 1918–28.

143. Henri Prêté, "L'action syndicale et la hausse des salaires," 1.

144. "Le Congrès de la Fédération des Travailleurs du Sous-sol: la question des salaires," 1.

145. "Le Congrès de la Fédération des Travailleurs du Sous-sol: la question des salaires," 2.

146. "Le Congrès de la Fédération des Travailleurs du Sous-sol: la question des salaires," 2.

147. Raynaud, *Vers le salaire minimum*, 224.

148. Statistique générale de la France, *Salaires et coût de l'existence à diverses époques, jusqu'en 1910*, 527.

149. Oualid and Picquenard, *La guerre et le travail*, 175, 267.

150. Oualid and Picquenard, *La guerre et le travail*, 175–76.

151. Oualid and Picquenard, *La guerre et le travail*, 187–91. See also Dutton, *Origins of the French Welfare State*, 16.

152. Oualid and Picquenard, *La guerre et le travail*, 224. Emphasis mine.

153. A.-M. Martinache, *"A travail égal, genre de vie égal,"* 29.

154. Roger Picard, *Le minimum légal de salaire*, 6.

155. Oualid and Picquenard, *La guerre et le travail*, 92.

156. Oualid and Picquenard, *La guerre et le travail*, 510.

157. Hyacinthe Dubreuil, *Robots or men?* 248.

158. Hyacinthe Dubreuil, "Pour la République industrielle," 1.

159. Dubreuil, "Pour la République industrielle," 2.

160. Dubreuil, "Pour la République industrielle," 2.

161. Jackie Clarke, "Engineering a New Order in the 1930s," 72.

162. Hyacinthe Dubreuil, *La chevalerie du travail*, 181.

163. Mary McLeod, "Urbanism and Utopia," 401; Eric Mumford, *The CIAM Discourse on Urbanism, 1928–1960*, 156.

164. Dubreuil, "Pour la République industrielle," 1.

CHAPTER SEVEN

1. François Perroux, *Science de l'homme et science économique*, 29.

2. Perroux, *Science de l'homme et science économique*, 29.

3. Perroux, *Science de l'homme et science économique*, 15.

4. Lebret et al., "Manifeste d'*Economie et Humanisme*," 2.

5. Archives Nationales[AN] AJ 72 1858. "'Assurez au pays son minimum vital' ordonne le président LAVAL aux préfets de zone occupée," *Aujourd'hui*.

6. AN AJ 72 1858. "Vite: le salaire minimum!" *Paris Soir*.

7. AN AJ 72 1858. "Une conquête ouvrière: le salaire minimum vital," *Inter-France*.

8. Jean Marchal, "Salaires et économie d'armistice," 122.

9. Olivier Dard, *Le rendez-vous manqué des relèves des années trente*, 332.

10. Amir Weiner, ed., *Landscaping the Human Garden*, 345. Mark Mazower, "Healthy Bodies, Sick Bodies," 76–103.

11. Andrés Horacio Reggiani, *God's Eugenicist*, 127.

12. Patrick Fridenson and Jean-Louis Robert, "Les ouvriers dans la France de la Seconde Guerre mondiale," 120.

13. Reggiani, *God's Eugenicist*, 114. Citing *Cahiers de la* Fondation *française de l'étude des problèmes humains* 4 (1945): 120.

14. Denis Pelletier, *Economie et Humanisme*, 275.

15. Reggiani, *God's Eugenicist*, 117.

16. Julian Jackson, "'Mal embarqué, bien arrivée,'" 157.

17. Lionel Robbins, *An Essay on the Nature and Significance of Economic Science*, 15.

18. Ivan Moscati, "Were Jevons, Menger, and Walras Really Cardinalists?" 301, 400.

19. Carl Menger, *Principles of Economics*, 52.

20. Menger, *Principles of Economics*, 77.

21. Moscati, "Were Jevons, Menger, and Walras Really Cardinalists?" 407.

22. Mark Blaug, *Economic Theory in Retrospect*, 570–82.

23. George J. Stigler and Gary S. Becker, "De Gustibus Non Est Disputandum," 76–90.

24. Knut Wicksell, *Lectures in Political Economy*, 43.

25. François Perroux, "Introduction," 32.

26. G. Destanne de Bernis, "François Perroux (1903–1987)," 494.

27. Perroux, *Science de l'homme et science économique*, 11. See also *Le pain et la parole* (Cerf, 1969): "Money must be deglorified." Cited in Destanne de Bernis, "François Perroux (1903–1987)," 495.

28. John Hellman, *The Knight-Monks of Vichy France*, 8. On Perroux's antirepublicanism see Jackson, "'Mal embarqué, bien arrivée,'" 160. Daniel Lindenberg, *Les années souterraines (1937–1947)*, 244.

29. Destanne de Bernis, "François Perroux (1903–1987)," 495.

30. Ludovic Frobert, "L'économie politique d'Étienne Antonelli (1879–1971)," 1529–59.

31. Olivier Marchand and Claude Thélot, *Deux siècles de travail en France*, 39. On p. 59 the authors give a lower figure of 1.2 million.

32. Robert Castel, *Les métamorphoses de la question sociale*, 324.

33. Jean-Louis Loubet del Bayle, *Les non-conformistes des années 30*, 562; Dard, *Le rendez-vous manqué des relèves des années trente*, 332.

34. See, for example, Jean Coutrot, *L'humanisme économique*, 145.

35. Loubet del Bayle, *Les non-conformistes des années 30*, 415. Citing René Dupuis and Alexandre Marc, *Jeune Europe* (Paris, 1933), 206.

36. Hellman, *The Knight-Monks of Vichy France*, 48.

37. Hellman, *The Knight-Monks of Vichy France*, 48; Jackson, "'Mal embarqué, bien arrivée,'" 163.

38. François Perroux, *Capitalisme et communauté de travail*, 346. Perroux collaborated with two other important corporatist theorists, Gaetan Pirou and Maurice Bouvier-Ajam. See Steven L. Kaplan, "Un laboratoire de la doctrine corporatiste sous le régime de Vichy," 35–77.

39. Jackson, "'Mal embarqué, bien arrivée,'" 160.

40. Perroux, "Introduction," 23, and *Science de l'homme et science économique*, 8.

41. Perroux, *Science de l'homme et science économique*, 13, 30.

42. François Perroux and Yves Urvoy, *Renaître, essais*; Perroux, *Science de l'homme et science économique*, 40, *Economie organisée et économie socialisée*, 23, and *Pour une politique des prix*, 15. In addition Perroux edited a series of short volumes on community and work for the Librairie de Médicis.

43. Lebret et al., "Manifeste d'*Economie et Humanisme*," 17. The economic sections of the first Economy and Humanism manifesto, written in November 1941 and published in the following year, were written by Perroux. They echo quite precisely his other pamphlets of the war years. Elsewhere the manifesto's more communitarian sections contain some elements from

which Perroux later distanced himself. He criticized Economie et Humanisme's founder, Father Louis Lebret, for promoting an impossible return to a precapitalist idyll. Instead Perroux called for new institutions appropriate to the modern economy. Other signatories of the manifesto, besides Lebret and Perroux, include Father Marie-Reginald Loew, agronomist Jean-Marius Gatheron, philosopher-peasant Gustave Thibon, industrialist Alexandre Dubois, Marie-Fabien Moos, and René Moreux. See Pelletier, *Economie et Humanisme*, 61.

44. Perroux, *Science de l'homme et science économique*, 31.

45. Robert Proctor, *Racial Hygiene: Medicine under the Nazis*, 415.

46. Perroux, *Science de l'homme et science économique*, 29.

47. Perroux, *Science de l'homme et science économique*, 33.

48. Perroux, *Science de l'homme et science économique*, 10.

49. Lebret et al., "Manifeste d'*Economie et Humanisme*," 9.

50. Perroux, *Science de l'homme et science économique*, 23.

51. Perroux, *Science de l'homme et science économique*, 10.

52. Perroux, *Science de l'homme et science économique*, 10n2. Filippo Carli again appears as an exemplar of this kind of synthesis, as does left-wing Swedish economist Gunnar Myrdal.

53. Perroux, *Science de l'homme et science économique*, 13.

54. Perroux, *Science de l'homme et science économique*, 14.

55. Perroux, *Science de l'homme et science économique*, 15.

56. Perroux, *Science de l'homme et science économique*, 31.

57. Reggiani, *God's Eugenicist*, 133. See also William H. Schneider, *Quality and Quantity*, 272–81.

58. Andrés Horacio Reggiani, "Alexis Carrel, the Unknown: Eugenics and Population Research under Vichy," 344. By contrast, Reggiani indicates, the entire Centre National de Recherches Scientifiques [CNRS] received 50 million francs in that same year.

59. Fondation française pour l'étude des problèmes humains, "Les premiers mois de travail de la Fondation française pour l'étude des problèmes humains," 18.

60. Reggiani, *God's Eugenicist*, 128.

61. Fondation française pour l'étude des problèmes humains, "Les premiers mois de travail de la Fondation française pour l'étude des problèmes humains," 43–44.

62. Reggiani, *God's Eugenicist*, 139.

63. See chap. 4 above.

64. Denis Peschanski, *Vichy 1940–1944: Contrôle et exclusion*, 209.

65. Reggiani, *God's Eugenicist*, 133.

66. Fondation française pour l'étude des problèmes humains, "Compte-rendu des activités de la Fondation française pour l'étude des problèmes humains pour l'année 1943," 11–117.

67. Fondation française pour l'étude des problèmes humains, "Compte-rendu des activités de la Fondation française pour l'étude des problèmes humains pour l'année 1943," 8.

68. Fondation française pour l'étude des problèmes humains, "Mémento de dépouillement pour l'enquête 'Régime Alimentaire.'" No date. Centre d'archives contemporaines [CAC] 760138 (5).

69. Fondation française pour l'étude des problèmes humains, "Enquête 'Régime alimentaire.'" CAC 760138 (4).

70. Fondation française pour l'étude des problèmes humains, "Enquête par sondage de l'état alimentaire d'une population" 9. ca. 1944. CAC 760138 (5).

71. Fondation française pour l'étude des problèmes humains, "Enquête par sondage de l'état alimentaire d'une population" 9. ca. 1944. CAC 760138 (5).

72. Perroux, *Science de l'homme et science économique*, 31.

73. Jean Forien de Rochesnard and Raymond Granier, "Le rationnement en France, 1939–1950," 4. AN AB XIX 4113.

74. Paul Reynaud, "Une arme de guerre, le rationnement," 19.

75. Dominique Veillon, *Vivre et survivre en France, 1939–1947*, 112.

76. "Message du Maréchal Pétain, chef d'Etat." August 13, 1940, and October 6, 1940. AN AB XIX 4033.

77. AN 72 AJ 1853. Emphasis mine.

78. Forien de Rochesnard and Granier, "Le rationnement en France, 1939–1950," 721.

79. Forien de Rochesnard and Granier, "Le rationnement en France, 1939–1950," 530.

80. Veillon, *Vivre et survivre en France, 1939–1947*, 231.

81. Joseph Weill, *Contribution à l'histoire des camps d'internement dans l'Anti-France*; Anne Grynberg, *Les camps de la honte: les internés juifs des camps français, 1939–1944*, 146-54; Renée Poznanski, *Les Juifs en France pendant la Seconde Guerre mondiale*, 226, 227, 266.

82. Juliette Lagabrielle, "Le rationnement alimentaire des enfants et des adolescents en France et en Europe," 35–45.

83. Randoin, *L'alimentation et la vie*, 11, 23, 91.

84. Randoin, *L'alimentation et la vie*, 7.

85. Charles Richet, "Études sur le rationnement alimentaire à l'Académie de Médecine," 72.

86. Pierre Chouard, "Note sur le Comité consultatif scientifique du Ravitaillement," 2–3. December 28. 1946. AN F23 380.

87. Arrêté du 28 octobre 1940. Cited in Jean Forien de Rochesnard and Raymond Granier, "Législation concernant le rationnement en France, 1934–1950," 73.

88. Richet, "Études sur le rationnement alimentaire à l'Académie de Médecine," 72.

89. Pierre Chouard, "Note sur le Comité consultatif scientifique du Ravitaillement" 1. AN F23 380.

90. See his papers at AN F23 379-401.

91. Forien de Rochesnard and Granier, "Le rationnement en France, 1939–1950," 298. Dozens of popular nutrition books and cookbooks claimed to guide housewives through the difficult task of meal planning. Titles published between 1941 and 1946 included the widely circulated *Cooking with Restrictions; Edible Insects; Manual for Living in Good Health;* and *Nourish Your Body: 100 Rational Recipes.*

92. Jean Forien de Rochesnard and Raymond Granier, "Législation concernant le rationnement en France, 1934–1950," 73.

93. Ministère de l'Agriculture et du Ravitaillement, Bureau Central des Recherches, *Cinquième rapport sur les travaux du Bureau Central des Recherches*, 92.

94. Lucie Randoin, *L'alimentation et la vie*, 91.

95. Lucie Randoin, "Les équilibres alimentaires," 117–53.

96. Bénitte, "Répercussions des restrictions alimentaires sur la santé publique," 41.

97. Bénitte, "Répercussions des restrictions alimentaires sur la santé publique," 42. AN F23 429.

98. Marchal, "Salaires et économie d'armistice," 91–92.

99. Richet, "Études sur le rationnement alimentaire à l'Académie de Médecine," 79.

100. Several such directors wrote to thank Jean Trémolières at the INH for having sent them monthly ration statistics. CAC 770621 (6).

101. A typed manuscript summarized the INH's wartime data, ca. 1945. See also P. Beyer, "Les différentes manifestations cliniques attribués en France à la carence alimentaire de 1940 à 1945," Recit des travaux (Institut National d'Hygiène). CAC 770621 (6).

102. Institut National d'Hygiène, "Données nutritionnelles sur l'alimentation des travailleurs" 6, ca. 1945. CAC 770621 (6).

103. Jean-François Picard, "The Institut national d'hygiène and Public Health in France, 1940–1946."

104. Archives of the Institut National de la Santé et la Recherche Médicale, previously the Institut National d'Hygiène. CAC 770621 (9-21). The Marseilles Center's director was Dr. Daniel Kuhlmann.

105. Centre d'études hygièniques de Marseille, "Etude sur l'état de nutrition de la population de Marseille" CAC 770621 (9).

106. Centre d'études hygièniques de Marseille, "Etude sur l'état de nutrition de la population de Marseille" CAC 770621 (9).

107. Institut National d'Hygiène, "Etude sur le rationnement." CAC 770621 (6).

108. Robert Menchérini, "Conséquences sanitaires des pénuries alimentaires dans les Bouches-du-Rhône," 419–32.

109. Institut National d'Hygiène, "Nutrition: Rations réellement consommées à Paris, à Marseille et dans une région rurale pendant le 2e semestre 1945." No date. AN F23 381.

110. Raymond Aubrac, Commisaire Régional de la République de la Région de Marseille, "Gravité exceptionnelle de la situation du ravitaillement Marseille." Letter to Ministre de Ravitaillement. No date.

111. Perroux, *Economie organisée et économie socialisée*, 20–21.

112. Charte du Travail, Article 54; Bozena Zabiecka, "Expériences de salaires minima à l'étranger et leurs repercussions économiques et sociales," 9, 25.

113. Miranda Pollard, *Reign of Virtue*, 151, citing René Belin, letter to prefects, September 28, 1940, AN F39 976. Translation is Pollard's.

114. Veillon, *Vivre et survivre en France, 1939–1947*, 52.

115. Dominique Veillon and Jean-Marie Flonneau, "Introduction," 98. Citing "1941, Bilan d'une année d'activité du Secours national dans la région parisienne," AN AB XIX 4120.

116. Menchérini, "Conséquences sanitaires des pénuries alimentaires dans les Bouches-du-Rhône," 422.

117. Veillon and Flonneau, "Introduction," 20.

118. Danièle Voldman, "Le logement: crise, pénurie, ou restrictions?" 381.

119. Arne Radtke, "La politique salariale de Vichy," 265n1.

120. *Prache*, *Gaston*, *Louis*, *Eugène*.

121. "Vite le salaire minimum!," *Paris Soir*.

122. "Une question d'actualité : le minimum d'existence," *Les Nouveaux Temps*.

123. "Une question d'actualité : le minimum d'existence," *Les Nouveaux Temps*, 1.

124. AN F22 1838. Robert Darrigol, Note relative à la Commission Supérieure des Salaires (February 24, 1944), 2.

125. AN F22 1511. Procès-verbal, Commission d'études relatives au coût de la vie, Préfecture du Loiret (June 16, 1941).

126. John F. Sweets, *Choices in Vichy France*, 17.

127. AN F22 1838. Conseil Supérieur du Travail, Centre d'Etude des Salaires, Etude no. 2: Calcul du salaire minimum vital (February 11, 1944), 2.

128. Centre d'Etude des Salaires, Etude no. 2. Calcul du salaire minimum vital, 5.

129. Centre d'Etude des Salaires, Etude no. 2. Calcul du salaire minimum vital, 3.

130. Eugène Schueller, *Du salaire-subsistance au salaire production*, 2.

131. Centre d'Etude des Salaires, Etude no. 2: Calcul du salaire minimum vital, 2.

132. Radtke, "La politique salariale de Vichy," 272.

133. Pollard, *Reign of Virtue*, 123–24.

134. Radtke, "La politique salariale de Vichy," 270, citing Circulaire Tr. 138 de René Belin aux préfets, August 11, 1941. AN F37 47.

135. Bozena Zabiecka, "Les salaires minima en France," 30.

136. Centre d'Etude des Salaires, Etude no. 2. Calcul du salaire minimum vital, 10.

137. Paul V. Dutton, *Origins of the French Welfare State*, 194.

138. AN AJ72 1858. Agence Française d'Informations de Presse, "Un problème qui n'a pas reçu sa solution définitive: Celui des salaires." No date.

139. AN AJ72 1858. Agence Française d'Informations de Presse, "Comment la notion du salaire minimum s'est substituée à celle du minimum vital." No date.

140. AN F22 1838. Raymond Deaud, Etude sur les besoins d'une famille ouvrière (1943), 16.

141. AN F22 1838. Raymond Deaud, Etude sur les besoins d'une famille ouvrière (1943), 15.

142. Jean Coutrot, *L'humanisme économique*, 61, 59. Coutrot took this slogan from Hyacinthe Dubreuil. On Coutrot, see Jackie Clarke, "Engineering a New Order in the 1930s," 63–86.

143. Perroux, *Science de l'homme et science économique*, 40.

144. Centre d'Etude des Salaires, Etude no. 2. Calcul du salaire minimum vital, 2.

145. Richard Kuisel, *Capitalism and the State in Modern France*, 344.

146. Institut National d'Etudes Démographiques, *Désirs des Français en matière d'habitation urbaine*.

147. Picard, "The Institut national d'hygiène and Public Health in France, 1940–1946."

148. Institut de Science Economique Appliquée, "Les caractères contemporains de la rémunération du travail," 175.

149. Perroux, *Economie organisée et économie socialisée*, 21.

150. Perroux, *Economie organisée et économie socialisée*, 4.

151. Institut de Science Economique Appliquée, "Salaire et consommation," 7.

152. Institut de Science Economique Appliquée, "Salaire et consommation," 7.

153. Institut de Science Economique Appliquée, "Salaire et consommation," 5.

154. Olivier Dard, "Théoriciens et praticiens de l'économie," 85.

155. Destanne de Bernis, "François Perroux (1903–1987)," 496.

156. Collège de France, "Professeurs disparus: François Perroux."

157. Perroux and his former colleague at Economy and Humanism Father Louis Lebret were key figures in the emerging field of development economics. The Economie et Humanisme group undertook dozens of studies on commission by Latin American governments. Economie et Humanisme created a model of development based on international expertise and a normative notion of human needs. Lebret was instrumental in the United Nations' development strategy based on a standard of "basic needs." Pelletier, *Economie et Humanisme*, 296, 357.

158. See, for example: "Rapport du Dr. Crosnier au Comité Consultatif Scientifique du Ravitaillement" April 9, 1945, AN F23 382; Ministère du Ravitaillement, "Calcul des besoins" July 19, 1945, AN F23 386; Ministère du Ravitaillement, "Compte tenu de la réunion relative au ravitaillement des détenus et internés politiques qui s'est tenue le mercredi 20 sept dans le bureau de M. Agard," 2, AN F23 386.

159. See chap. 4, "Social Justice and the Provisioning Crisis," in Megan Koreman, *The Expectation of Justice*, 340.

CHAPTER EIGHT

1. Centre d'archives contemporaines [CAC] 760121 153. Procès-verbal de la Sous-commission du Budget-type (partie non-alimentaire): mardi 23 mai 1950 (May 23, 1950), 6.

2. CAC 760121 153. Procès-verbal de la Sous-commission du Budget-type (partie non-alimentaire): mardi 23 mai 1950 (May 23, 1950), 6.

3. Dominique Veillon, *Vivre et survivre en France, 1939–1947*, 200.

4. CAC 820774 (24). Délégation Générale à l'Equipement National, Enquête sur l'équipement du logement (ca. 1944).

5. Megan Koreman, *The Expectation of Justice: France, 1944–1946*, 3.

6. Procès-verbal de la Sous-commission du Budget-type (partie non-alimentaire): mardi 23 mai 1950, 6.

7. This was the well-known conclusion of sociologist Maurice Halbwachs's 1933 study of working-class consumption. Maurice Halbwachs, *L'évolution des besoins dans les classes ouvrières*, 96.

8. Claude Lapierre, "L'élaboration du budget-type et la fixation du salaire minimum garanti," 382.

9. CAC 750121 (153). Procès-verbal de la Sous-commission du Budget-type (partie non-alimentaire) du 16 mai 1950 (May 16, 1950).

10. Victoria De Grazia, *Irresistible Empire*, 101–2. See also Judith Coffin, "A Standard of Living?" 7; Daniel Horowitz, *The Morality of Spending*, 254.

11. Konrad A. Jarausch and Michael Geyer, "In Pursuit of Happiness," 275–86; De Grazia, *Irresistible Empire* 336–75.

12. Paul V. Dutton, *Origins of the French Welfare State*, 210.

13. Stephen Marglin, "Lessons of the Golden Age," 15.

14. Armin Triebel, "Moral und Ökonomie," 384–92.

15. CAC 760121 (153). Commission Supérieure des Conventions Collectives (April 1950).

16. CAC 760121 (153). Commission Supérieure des Conventions Collectives: Procès-verbal de la séance tenue le 8 mai 1950, 3.

17. CAC 760121 (153). Commission Supérieure des Conventions Collectives: Procès-verbal de la séance tenue le 8 mai 1950, 4.

18. R. Bothereau, "Beaucoup de bruit pour rien."

19. CAC 760121 (153). Commission Supérieure des Conventions Collectives: Procès-verbal de la séance tenue le 8 mai 1950, 5.

20. CAC 706121 (153). Commission Supérieure des Conventions Collectives: Procès-verbal de la séance tenue le 12 mai 1950, 3.

21. Confédération Française Démocratique du Travail archives [CFDT]e 4 H 96. Confédération Générale du Travail, Centre Confédéral d'Etudes Economiques et Sociales: Commission du coût de la vie et du minimum vital (ca. 1947), 1.

22. Archive Nationales [AN] AJ 72 1858. Agence Française d'Information de Presse, "Comment la notion du salaire minimum s'est substituée à celle du minimum vital" (ca. 1943).

23. Y. Séquillon, *L'évolution des salaires*, 1.

24. CFDT 4 H 96. Confédération Française des Travailleurs Chrétiens, Etude: Observations sur le minimum vital établi par la C.F.T.C. (No date..)

25. Jean-Pierre Rioux, *La France de la Quatrième République*, 173.

26. Koreman, *The Expectation of Justice*, 54.

27. Rioux, *La France de la Quatrième République*, 120.

28. Philippe Buton, "L'éviction des ministres communistes," 340.

29. Koreman, *The Expectation of Justice*, 156.

30. Jean-Louis Guglielmi and Marguerite Perrot, *Salaires et revendications sociales en France, 1944–1952*, 31.

31. CAC 760121 (153). Commission Supérieure des Conventions Collectives: Procès-verbal de la séance tenue le 8 mai 1950, 3.

32. AN 72 AJ 546. H. Bernard, Secrétairat des Commissions d'Etudes des Problèmes de l'Après-guerre. Commission économique financière et sociale. Section sociale. Salaire minimum (June 15, 1943), 4–5. See also Dutton, *Origins of the French Welfare State*, 204–7.

33. AN F22 2030. Comité National de Resistance, Un programme d'action du C.N.R. (March 15, 1944), 4.

34. Rioux, *La France de la Quatrième République*, 179.

35. Guglielmi and Perrot, *Salaires et revendications sociales en France, 1944–1952*, 70.

36. Claude Lapierre, "Les accords de salaires," 15–22.

37. Alain Bayet, "L'accroissement spectaculaire des salaires et leur pouvoir d'achat," 170.

38. The so-called Commission Délépine included Henri Raynaud and André Barjonet (CGT) and Gustave Salmon (CFTC), all active participants in the SMIG commission.

39. CFDT 4 H 96. Conseil Supérieure de la Fonction Publique, Rapport de la Commission du minimum vital (n.d.), 1.

40. CAC 760121 (153). Commission Supérieure des Conventions Collectives: Procès-verbal de la séance tenue le 8 mai 1950, 6.

41. Susan Pedersen, *Family, Dependence, and the Origins of the Welfare State*, 478.

42. Marilyn Boxer, "Protective Legislation and the Marginalization of Women Workers," 45–66.

43. Dutton, *Origins of the French Welfare State*, 28.

44. CAC 760121 (153). Commission Supérieure des Conventions Collectives: Procès-verbal de la séance tenue le 8 mai 1950, 7.

45. Judith Coffin, "Social Science Meets Sweated Labor," 261.

46. AN 72 AJ 546. Leroy, Salaire minimum et salaire féminin (June 18, 1943), 3–5. France Combattante. Commission économique financière et sociale. Section sociale.

47. CAC 760121 (283). Direction du Travail; Sous-direction des Conditions du Travail, Lettre au Ministre des Affaires Etrangères (November 30, 1955), 3.

48. CFDT 4 H 106. Délépine, Rapport, Sous-commission du minimum vital (February 20, 1947), 9.

49. CAC 760121 (153). Commission Supérieure des Conventions Collectives: texte adopté au cours de la séance du 12 mai 1950.

50. CAC 760121 (153). Sous-commission chargée de l'étude de la partie alimentaire du budget-type, procès-verbal de la séance du 16 mai 1950, 3.

51. CAC 760121 (153). Sous-commission chargée de l'étude de la partie alimentaire du budget-type, procès-verbal de la séance du 16 mai 1950, 3.

52. CFDT 4 H 96. G. Villiers (CNPF), Lettre au Ministre du Travail et de la Sécurité Sociale (May 31, 1950), 1.

53. CFDT 4 H 96. G. Villiers (CNPF), Lettre au Ministre du Travail et de la Sécurité Sociale (May 31, 1950), 1.

54. Société d'Etudes Economiques et Documentaires, "Aperçus véridiques sur la farce du minimum interprofessionnel," 5.

55. CAC 760121 (282). Raymond Lévy-Bruhl, Note pour M. le Ministre; Objet: budget-type (November 21, 1952), 1.

56. CAC 760121 (282). Raymond Lévy-Bruhl, Note pour M. le Ministre; Objet: budget-type (November 21, 1952), 1.

57. CAC 760121 (283). F. Bardin, le Médecin Controlleur Général. Note pour M. le Chef de Service du Contrôle Général de la Sécurité Sociale (June 27, 1955).

58. Direction du Travail; Sous-direction des Conditions du Travail, Lettre au Ministre des Affaires Etrangères, 4.

59. Union Nationale des Caisses d'Allocations Familiales, *Les causes matérielles de la dénatalité*, 18.

60. CFDT 4 H 96. André Barjonet, C.G.T. Centre Confédéral d'Etudes Economiques et Sociales; Commission du Coût de la Vie et du Minimum Vital (n.d.), 1.

61. CAC 760121 (153). Documents pour le travail de la sous-commission de la partie alimentaire, préparés pour M. Houist (1950).

62. CFDT 4 H 96. G. Villiers (CNPF), Lettre au Ministre du Travail et de la Sécurité Sociale (May 31, 1950), 2.

63. CFDT 4 H 106. Délépine, Rapport: La Sous-commission du Minimum vital (20 February 1947) 2.

64. AN F23 (380). Lucie Randoin, "Considérations rélatives à la valeur calorifique et à la teneur en protides et en lipides de la ration alimentaire actuelle du parisien adulte" (1946).

65. CAC 760121 (153). Commission Supérieure des Conventions Collectives: Procès-verbal de la séance tenue le 18 juillet 1950 4.

66. CFDT 4 H 96. André Barjonet, C.G.T. Centre Confédéral d'Etudes Economiques et Sociales; Commission du Coût de la Vie et du Minimum Vital (n.d., c.a. 1947) 1.

67. CFDT 4 H 106. Délépine, Rapport: La Sous-commission du Minimum vital, 4.

68. Jean Trémolières and F. Vinit, "Enquête sur l'état de nutrition de la population, Paris (hiver 1950–1951)," 642.

69. Jean Trémolières, F. Vinit, and Y. Serville, "Etude sur la ration alimentaire type à préconiser pour le Français," 768.

70. Trémolières, Vinit, and Serville, "Etude sur la ration alimentaire type à préconiser pour le Français," 780.

71. Trémolières, Vinit, and Serville, "Etude sur la ration alimentaire type à préconiser pour le Français," 787.

72. Eugène Bellut, "Rapport présenté au nom de la sous-commission du budget alimentaire," 16.

73. CAC 760121 (154). Commission Supérieure des Conventions Collectives, Sous-commission chargée de l'étude de la partie alimentaire: procès-verbal de la 4ème séance tenu le 16 Octobre 1953, 11.

74. CAC 760121 (282). Raymond Lévy-Bruhl, Note pour M. le Ministre; Objet: budget-type (November 21, 1952), 2.

75. CAC 760121 (154). Commission Supérieure des Conventions Collectives, Sous-commission chargée de l'étude de la partie alimentaire: procès-verbal de la 4ème séance tenu le 16 Octobre 1953, 4.

76. Comment by Dr. Jacquot on Lucie Randoin et al., "Méthodes et résultats des enquêtes sur la consommation alimentaire française d'avant-guerre (1937–1938–1939)," in *Les enquêtes de consommation en France*, 54.

77. CAC 760121 (154). Commission Supérieure des Conventions Collectives, Sous-commission chargée de l'étude de la partie alimentaire: procès-verbal de la 4ème séance tenu le 16 octobre 1953, 2.

78. CFDT 4 H 106. Délépine, Rapport: La Sous-commission du Minimum vital (February 20, 1947), 4.

79. AN F23 381. Institut National d'Hygiène, Rations réellement consommées à Paris, à Marseille et dans des régions rurales pendant l'année 1946 par individu moyen et par jour (n.d., ca. 1947).

80. CAC 760121 (153). Sous-commission chargée de l'étude de la partie alimentaire du budget-type, procès-verbal de la séance du 19 mai 1950, 2.

81. CAC 760121 (153). Sous-commission chargée de l'étude de la partie alimentaire du budget-type, procès-verbal de la séance du 19 mai 1950, 2.

82. CAC 760121 (153). Sous-commission chargée de l'étude de la partie alimentaire du budget-type, procès-verbal de la séance du 31 mai 1950, 6.

83. CFDT 4 H 106. Délépine, Rapport: La Sous-commission du Minimum vital (February 20, 1947) 9.

84. André Barjonet, "Une revendication fondamentale: celle du minimum vital," 6.

85. CFDT 4 H 106. Conseil National du Patronat Français, Les calculs du Budget dit 'Minimum vital . . .' (n.d.).

86. Dutton, *Origins of the French Welfare State*, 210.

87. See chap. 6, "Wartime France as a Human Laboratory" in Andrés Horacio Reggiani, *God's Eugenicist*, 268.

88. Johan Heilbron, "Pionniers par défaut?" 365–79; Peter Miller and Nikolas Rose, "Governing Economic Life," 1–31.

89. *Les enquêtes de consommation en France*, ed. J.-L. Dols, 157.

90. Heilbron, "Pionniers par défaut?" 370.

91. Michel Amiot, *Contre l'état, les sociologues*, 36.

92. John Hellman, *The Knight-Monks of Vichy France*, 123.

93. Paul Henry Chombart de Lauwe, *La vie quotidiennne des familles ouvrières*, 148.

94. Chombart de Lauwe, *La vie quotidiennne des familles ouvrières*, 148.

95. Paul Henry Chombart de Lauwe, *Paris et l'agglomération parisienne*.

96. Amiot, *Contre l'état, les sociologues*, 38.

97. Chombart de Lauwe, *La vie quotidiennne des familles ouvrières*, 149.

98. Chombart de Lauwe, *La vie quotidiennne des familles ouvrières*, 150.

99. Chombart de Lauwe, *La vie quotidiennne des familles ouvrières*, 163.

100. Chombart de Lauwe, *La vie quotidiennne des familles ouvrières*, 232.

101. Chombart de Lauwe, *La vie quotidiennne des familles ouvrières*, 94–97.

102. Raymond Lévy-Bruhl, "Méthodes d'enquête auprès des familles et quelques cas d'application étudiés par l'INSEE," 107.

103. Amiot, *Contre l'état, les sociologues*, 40.

104. CFDT 4 H 96. André Barjonet, CGT Centre Confédéral d'Etudes Economiques et Sociales; Commission du coût de la vie et du minimum vital, 2.

105. CAC 750121 (153). Procès-verbal de la Sous-commission du Budget-type (partie non-alimentaire): mercredi 24 mai 1950.

106. Conseil Supérieur de la Fonction Publique; Rapport de la Commission du Minimum Vital, 4.

107. CFDT 4 H 106. Délépine, Rapport: La Sous-commission du Minimum vital, 6.

108. Chombart de Lauwe, *La vie quotidiennne des familles ouvrières*, 190.

109. Lapierre, "L'élaboration du budget-type et la fixation du salaire minimum garanti," 382.

110. Barjonet, "Une revendication fondamentale: celle du minimum vital," 6.

111. CAC 750121 (153). Procès-verbal de la Sous-commission du Budget-type (partie non-alimentaire): mercredi 31 mai 1950, 6–7.

112. CAC 750121 (153). Procès-verbal de la Sous-commission du Budget-type (partie non-alimentaire): mardi 30 mai 1950, 7–8.

113. AN 72 AJ 546. P. Chauvet, Rapport sur l'établissement d'un salaire minimum et sa liaison avec le sursalaire familial (1942), 1.

114. CFDT 4 H 106. Délépine, Rapport: La Sous-commission du Minimum Vital, 10.

115. CFDT 4 H 96. Conseil Supérieur de la Fonction Publique; Rapport de la Commission du Minimum Vital (n.d., c.a. 1947), 5.

116. CAC 760121 (282). Raymond Lévy-Bruhl, Note pour M. le Ministre; Objet: budget-type (November 21, 1952).

117. Lapierre, "L'élaboration du budget-type et la fixation du salaire minimum garanti," 382.

118. CAC 760121 (283). Direction du Travail: Sous-direction des Conditions du Travail (December 23, 1955).

119. "It was in 1946 that the CGT launched the idea of a minimum vital, which was then copied by the other unions." André Barjonet, "Ce que vous devez savoir sur le minimum vital," 4.

120. L. Mascarello, "Une insulte à la classe ouvrière!" 1–3.

121. CFDT 8 H 844. Confédération Française des Travailleurs Chrétiens, Circulaire aux Unions départementales et Fédérations de Métier (September 16, 1950), 5.

122. Habitus, or social practice, constitutes one of the aspects of class delineated in recent scholarship, alongside identity and social position. Jan E. Goldstein, "Of Marksmanship and Marx," 89, 106; Neville Kirk, "Decline and Fall, Resilience and Regeneration," 94.

123. Evelyne Bughin and Jean-François Payen, "Le salaire horaire minimum depuis 1950," 22.

124. Peter Miller and Ted O'Leary, "Accounting and the Construction of the Governable Person," 258.

CHAPTER NINE

1. Raymond Williams, "Ideas of Nature," 76.

2. Ramachandra Guha, *How Much Should People Consume?* 6.

3. Robert Fogel, *The Escape from Hunger and Premature Death, 1700–2100*, 27.

4. Fogel, *The Escape from Hunger and Premature Death*, 33.

5. Fogel, *The Escape from Hunger and Premature Death*, 37.

6. Fogel, *The Escape from Hunger and Premature Death*, 34.

7. Fogel, *The Escape from Hunger and Premature Death*, xv.

8. Fogel, *The Escape from Hunger and Premature Death*, 59.

9. Joseph Degerando, "Considérations dans les diverses méthodes à suivre dans l'observation des peuples sauvages," 155.

10. Christopher Hamlin, "The 'Necessaries of Life' in British Political Medicine, 1750–1850," 379.

11. Kate Soper, "Conceptualizing Needs in the Context of Consumer Politics," 356.

12. These questions reflect the very modern problem of how to quantify human life. Theodore M. Porter, "Objectivity as Standardization," 210–20.

13. E. J. Hobsbawm, "The British Standard of Living, 1790–1850," 47, 50.

14. E. J. Hobsbawm, "The Standard of Living during the Industrial Revolution," 131.

15. Stanley L. Engerman, "Reflections on 'the Standard of Living Debate,'" 51.

16. John Komlos, "Preface," xi–xiii; Clayne L. Pope, "The Changing View of the Standard-of-Living Question in the United States," 334.

17. Wally Seccombe, *Weathering the Storm*, 73.

18. Engerman, "Reflections on 'the Standard of Living Debate,'" 52.

19. Sonja Lyubomirsky, *The How of Happiness*, 260.

20. Amartya Sen, *Inequality Reexamined*, 28.

21. Sudhir Anand and Amartya K. Sen, *Sustainable Human Development*, 58.

22. Walter Gropius, "Sociological Premises for the Minimum Dwelling of Urban Industrial Populations," 113.

23. Karel Teige, *The Minimum Dwelling*, 20.

24. Christina Kaier, "The Russian Constructivist Flapper Dress," 188.

25. Herbert Marcuse, *One-Dimensional Man*, 4.

26. Ronald Inglehart and Jacques René Rabier, "Political Realignment in Advanced Industrial Society," 456–79.

27. W Steffen et al., "The Anthropocene," 849–53.

28. Guha, *How Much Should People Consume?* 233.

29. Kate Soper, *What Is Nature?* 168.

30. Kate Soper, "The Politics of Nature," 66, and "Conceptualizing Needs in the Context of Consumer Politics," 356.

BIBLIOGRAPHY

ARCHIVES

Archives Nationales [AN]
Centre d'archives contemporaines [CAC]
Confédération française des travailleurs [CFDT]
Service historique des armées de terre [SHAT]

PRINTED SOURCES

Académie des Sciences Morales et Politiques. "Emile Cheysson." http://www.asmp.fr/fiches_academiciens/decede/CHEYSSON.htm (accessed November 21, 2011).

Administration générale de l'assistance publique à Paris. *Régime alimentaire: tableau des denrées d'assaisonnement*. Paris: Paul Dupont, 1851.

Agamben, Giorgio. *Homo Sacer: Sovereign Power and Bare Life*. Translated by Daniel Heller-Roazen. Edited by Werner Hamacher and David E. Wellbery. Meridian: Crossing Aesthetics. Stanford, CA: Stanford University Press, 1998.

Agulhon, Maurice. 1848, *ou l'apprentissage de la République: nouvelle histoire de la France contemporaine*. Paris: Seuil, 1973.

Aisenberg, Andrew R. *Contagion: Disease, Government, and the "Social Question" in Nineteenth-Century France*. Stanford, CA: Stanford University Press, 1999.

A.L. *Les bouches inutiles*. Paris: A. Lainé, 1870.

Alder, Ken. "A Revolution to Measure: The Political Economy of the Metric System in France." In *The Values of Precision*, edited by M. Norton Wise, 39–71. Princeton, NJ: Princeton University Press, 1995.

Amiot, Michel. *Contre l'état, les sociologues: éléments pour une histoire de la sociologie urbaine en France (1900–1980)*. Paris: Editions de l'EHESS, 1986.

Anand, Sudhir, and Amartya K. Sen. *Sustainable Human Development: Concepts and Priorities*. New York: United Nations, 1996.

Appert, Benjamin. *Bagnes, prisons et criminels*. Second ed. 4 vols. Paris: Guilbert, 1836.

Argemí i d'Abadal, Lluis. "Agriculture, Agronomy, and Political Economy: Some Missing Links." *History of Political Economy* 34, no. 2 (2002): 449–78.

Asselin, Jean-Charles. "La stagnation économique." In *Entre l'état et le marché: l'économie français de 1880 à nos jours*, edited by Maurice Lévy-Leboyer and Jean-Claude Casanova, 199–250. Paris: Gallimard, 1991.

" 'Assurez au pays son minimum vital' ordonne le président LAVAL aux préfets de zone occupée." *Aujourd'hui*, January 22, 1943. AN AJ 72 1858.

Bange, Christian, Pietro Corsi, and Pascal Duris. "Les médecins auditeurs du cours professé par Lamarck au Muséum (1795–1823): premier bilan d'une recherche." *Histoire des Sciences Médicales* 24, no. 4 (2000): 395–407.

Bardet, G. *La ration normale: moyens pratiques de l'établir*. Paris: Octave Doin & Fils, 1917.

Barjonet, André. "Une revendication fondamentale: celle du minimum vital." *Le Peuple* 29, no. 227 (1948): 6.

———. "Ce que vous devez savoir sur le minimum vital." *Le Peuple* 31, no. 309 (1950): 4.

Barzman, John. "La 'crise de la vie chère' de l'été 1919 au Havre." *Le Mouvement social* 170 (1995): 61–84.

Baudelot, Christian, and Roger Establet. *Maurice Halbwachs: consommation et société*. Paris: Presses Universitaires de France, 1994.

Baudrillart, Henri. "Le luxe et les moralistes: la théorie du retranchement des besoins." *Journal des Economistes*, 3rd series, 11, no. 40 (1876): 355–77.

———. *Philosophie de l'économie politique: des rapports de l'économie politique et de la morale*. Second ed. Economistes et publicistes contemporains. Paris: Gillaumin, 1883.

Bayet, Alain. "L'accroissement spectaculaire des salaires et leur pouvoir d'achat." In *Le travail en France, 1800–2000*, edited by Olivier Marchand and Claude Thélot, 155–72. Paris: Nathan, 1997.

Becquerel, Alfred. *Traité élémentaire d'hygiène privée et public*. Second ed. Paris: Labé, Librairie de la faculté de médecine, 1854.

Bellut, Eugène. "Rapport présenté au nom de la sous-commission du budget alimentaire." *Revue française du travail* 9, no. 1 (1954): 12–44.

Bénitte. "Répercussions des restrictions alimentaires sur la santé publique." In *Cinquième rapport sur les travaux du Bureau Central des Recherches, Annexe III*, edited by Bureau Central des Recherches, Ministère de l'Agriculture de du Ravitaillement, 41–47. Paris: Imprimerie Nationale, 1942.

Bensaude-Vincent, Bernadette. "The Balance: Between Chemistry and Politics." *The Eighteenth Century: Theory and Interpretation* 33, no. 3 (1992): 217–37.

Bérard, Pierre Honoré. *Cours de physiologie, fait à la Faculté de médecine de Paris*. Vols. 1–3, Paris: Labé, éditeur, Libraire de la faculté de médecine, 1848–51.

———. "Rapport sur le régime alimentaire des lycées de Paris." In *Dictionnaire d'hygiène publique et de salubrité*, edited by Ambroise Tardieu, 334–43. Paris: J.-B. Ballière, 1854.

Berg, Maxine. *The Machinery Question and the Making of Political Economy*, 1815–1848. Cambridge: Cambridge University Press, 1980.

Bernardi, Peter J. Maurice *Blondel, Social Catholicism, and Action Française: The Clash over the Church's Role in Society during the Modernist Era*. Washington, DC: Catholic University of America Press, 2009.

Berriat-Saint-Prix, Charles. *Mazas: Etude sur l'emprisonnement individuel.* Paris: Cosse et Marechal, 1860.

Blanc, Louis. *Révolution française: histoire de dix ans, 1830–1840*. Vol. 4. Paris: Pagnerre, 1843.

———. *Organisation du Travail.* Fifth ed. Paris: Société de l'Industrie Fraternelle, 1847.

———. *La révolution de février au Luxembourg*. Paris: Michel Lévy, 1849.

Blanqui, Auguste. "Le rationnement." *La Patrie en Danger*, December 8, 1870, 1.

Blaug, Mark. *Economic Theory in Retrospect*. Fifth ed. Cambridge: Cambridge University Press, 1997.

Bonin, Hubert. 1998. "Employment and the Revolution of 1848 in France." http://www.ohiou.edu/~Chastain/dh/francemp.htm (accessed November 5, 2003).

Bonvoisin, Georges. *Allocations familiales et caisses de compensation: leur origine, leur raison d'être, leurs effets, leur fonctionnement dans la Région Parisienne*. Paris: Lang, Blanchong et cie., 1927.

Bordarier, Paul. "Le siège de Paris d' après un recueil de lettres-ballons." *La Revue Hommes et Mondes* 10, no. 39 (1949): 255–75.

Bothereau, R. "Beaucoup de bruit pour rien." *Le Peuple*, December 28 1946, 1.

Bouge, Auguste. "Les conditions du travail et le collectivisme." *Revue Politique et Parlementaire* 9 (1896): 241–97.

Boulaine, Jean, and Jean-Paul Legros. "Gasparin." In *D'Olivier de Serres à René Dumont, portraits d'agronomes*, 53–66. London: Technique & Documentation, 1998.

Bourgoin, Edme. *Hygiène publique: de l'alimentation des enfants et des adultes dans une ville assiégée et en particulier de la viande de cheval*. Paris: Adrien Delahaye, 1870.

Bourguinat, Nicolas. *Les grains du désordre: l'état face aux violences frumentaires dans la première moitié du XIXe siècle*. Paris: Editions de l'EHESS, 2002.

Boussingault, Jean Baptiste. "Analyses comparées des alimens consommés et des produits rendus par une vache laitière; recherches entreprises dans le but d'examiner si les animaux herbivores empruntent de l'azote à l'atmosphère." *Annales de chimie et de physique* 72 (1839): 113–27.

———. *Economie rurale considérée dans ses rapports avec la chimie, la physique, et la météorologie*. First ed., 2 vols. Vol. 2. Paris: Béchet Jeune, 1844.

———. "Recherches sur la quantité d'acide carbonique contenue dans l'air de la ville de Paris." *Annales de chimie et de physique*, 3rd series, no. 10 (1844): 456–74.

———. "Analyses comparées de l'aliment consommé et des excréments rendus par une tourterelle, entreprises pour rechercher s'il y a exhalation d'azote pendant la respiration des granivores." *Annales de chimie et de physique*, 3rd series, no. 11 (1844): 433–56.

———. "Recherches expérimentales sur le développement de la graisse pendant l'alimentation des animaux." *Annales de chimie et de physique*, 3rd series, no. 14 (1845): 419–82.

———. *Economie rurale considérée dans ses rapports avec la chimie, la physique et la météorologie*. Second ed, 2 vols. Vol. 2, Paris: Béchet Jeune, 1851.

———. *Mémoires de J.-B. Boussingault*. Vol. 4. Paris: Chamerot et Renouard, 1903.

Boxer, Marilyn. "Protective Legislation and the Marginalization of Women Workers." *Journal of Social History* 20, no. 1 (1986): 45–66.

Bresson, F. "La question de la viande." *Le Figaro*, October 28, 1870, 3.

Breton, Yves, and Michel Lutfalla, eds. *L'économie politique en France au XIXe siècle*. Paris: Economica, 1991.

"Budgets et salaires." *L'Information ouvrière et sociale*, July 20, 1919, 6.

Bughin, Evelyne, and Jean-François Payen. "Le salaire horaire minimum depuis 1950: pouvoir d'achat et influence sur l'eventail des salaires ouvriers." *Problèmes économiques* 1920 (1985): 21–26.

Buret, Eugène. *La misère des clases laborieuses en Angleterre et en France; de la nature de la misère, de son existence, des ses effets, de ses causes, et de l'insuffisance des remèdes qu'on lui a opposés jusqu'ici; avec l'indication des moyens propres à en affranchir les sociétés*. 2 vols. Vol. 1. Paris and Liepzig: Jules Renouard, 1841.

Buton, Philippe. "L'éviction des ministres communistes." In *L'Année* 1947, edited by Serge Berstein and Pierre Milza, 339–56. Paris: Presses de Sciences Po, 2000.

Cabanis, Pierre-Jean-Georges. *Rapports du physique et du moral de l'homme*. Eighth ed. Paris: Ballière, 1844.

Cacheux, Emile. *L'économiste pratique: construction et organisation des crèches, salles d'asile, écoles, habitations ouvrières et maisons d'employés, hôtels pour célibataires, cuisines économiques, bains, lavoirs, cercles populaires, nourriciers, maternités, dispensaires, hôpitaux, hospices, asiles de nuit, postes de sécours; mécanisme, statuts, règlements des institutions de prévoyance et de bienfaisance*. Paris: Librairie polytéchnique Baudry, 1885.

Carpenter, Kenneth. *Protein and Energy: A Study of Changing Ideas in Nutrition*. Cambridge: Cambridge University Press, 1994.

Castel, Robert. *Les métamorphoses de la question sociale: un chronique du salariat*. Edited by Pierre Birnbaum. L'espace du politique. Paris: Fayard, 1995.

Chaigneau, Marcel. *Jean-Baptiste Dumas: sa vie, son oeuvre, 1800–1884*. Paris: Guy Le Prat, 1984.

Chappey, Jean-Luc. *La Société des Observateurs de l'Homme (1799–1804): des anthropologues au temps de Bonaparte*. Paris: Société des études robespierristes, 2002.

Chaptal, Jean-Antoine. *Chimie appliquée à l'agriculture*. Vol. 1. Paris: Madame Huzard, 1823.

Charles, A. *Législation des établissements publics d'instruction secondaire*. Paris: Imprimerie et librairie classiques, Jules Delalain et fils, 1872.

Chassevant, Allyre. *Le régime alimentaire de guerre*. Collection de l'Action Nationale. Paris: Bureaux de l'Action Nationale, 1918.

Cheysson, Emile. *Société anonyme de Passy-Auteuil pour les habitations ouvrières, Rapport à l'assemblée générale du 22 juin 1881*. Extract from *l'Economiste français*, August 27, 1881. Paris: Imprimerie Chaix, 1881.

———. *Le salaire au point de vue statistique, économique et social*. Extract from the *Bulletin de la Société d'économie sociale*. Paris: Bureaux de la Réforme Sociale, 1884.

———. "La loi de l'hérédité." *Le Temps*, October 23 1885, 3.

———. *La question de la dépopulation en France. Revue Politique et Parlementaire*, October 1896. Paris: Bureaux de la Revue Politique et Parlementaire, 1896.

———. *Frédéric Le Play, sa doctrine, sa méthode, son école: communication faite à l'Académie des Science Morales et Politiques*. Paris: Guillaume, 1905.

Cheysson, Emile, and Alfred Toqué. *Les budgets comparés des cent monographies de familles*. 1890 ed. Geneva: Slatkine Reprints, 1984.

Chombart de Lauwe, Paul Henry. *Paris et l'agglomération parisienne*. 2 vols. Paris: Presses Universitaires de France, 1952.

———. *La vie quotidiennne des familles ouvrières (recherches sur les comportements de consommation): Travaux du Centre d'Études Sociologiques*. Paris: CNRS, 1956.

Clamageran, Jean-Jules. *Souvenirs du siège de Paris: cinq mois à l'Hôtel de Ville* (septembre 1870 janvier 1871). Paris: Gillaumin, 1872.

Clarke, Jackie. "Engineering a New Order in the 1930s: The Case of Jean Coutrot." *French Historical Studies* 24, no. 1 (2001): 63–86.

Clayson, Hollis. *Paris in Despair: Art and Everyday Life under Siege* (1870–71). Chicago: University of Chicago Press, 2002.

Coffin, Judith. "Social Science Meets Sweated Labor: Reinterpreting Women's Work in Late Nineteenth-Century France." *Journal of Modern History* 63, no. 2 (1991): 230–270.

———. "Consumption, Production, and Gender: The Sewing Machine in Nineteenth-Century France." In *Gender and Class in Modern Europe*, edited by Laura L. Frader and Sonya O. Rose, 111–41. Ithaca, NY: Cornell University Press, 1996.

———. "A Standard of Living? European Perspectives on Class and Consumption in the Early Twentieth Century." *International Labor and Working-Class History* 55 (1999): 6–26.

Cole, Joshua. *The Power of Large Numbers: Population, Politics, and Gender in Nineteenth-Century France*. Ithaca, NY: Cornell University Press, 2000.

Coleman, William. *Death Is a Social Disease: Public Health and Political Economy in Early Industrial France*. Madison: University of Wisconsin Press, 1982.

Collège de France. 2004. "Professeurs disparus: François Perroux." http://www.college-de-france.fr/default/EN/all/ins_dis/p1100090437010.htm (accessed June 2007).

Commission des logements insalubres, Département de la Seine, Ville de Paris. *Rapports généraux sur les travaux de la Commission pendant les années 1851 à 1869*. Paris: Charles de Mourgues Frères, 1877.

Congrès International de statistique. "Résolutions adoptées à Bruxelles. Budget économique des classes laborieuses. Dépenses habituelles de première nécessité. Dépenses improductives de luxe. Manière de rendre les resultats comparables." In *Compte-rendu général des travaux du Congrès International de statistique dans les séances tenus à Bruxelles 1853, Paris 1855, Vienne 1857, et Londres 1860*, edited by Ernst Engel, 14–16. Berlin: Imprimerie Royale (R. Decker), 1860.

Conseil de salubrité. "Instruction concernant les moyens d'assurer la salubrité des habitations." *Ordonnances et arrêtés du Préfet de Police* (1848): 611–18.

Conseil de santé des armées. *Instruction à l'effet de guider les troupes dans la composition de leur régime alimentaire* (5 mars 1850). Strasbourg: Librairie militaire de V. Berger-Levrault, 1850.

Conseil Supérieur du Travail. *Septième session (Décembre 1897)*. Edited by Ministère du Commerce, de l'Industrie, des Postes et des Télégraphes. Paris: Imprimerie nationale, 1897.

Cooper, Frederick, and Randall M. Packard. "Introduction." In *International Development and the Social Sciences: Essays on the History and Politics of Knowledge*, edited by Frederick Cooper and Randall M. Packard, 1–44. Berkeley: University of California Press, 1997.

Coulomb, Charles Augustin. "Résultat de plusieurs expériences destinées à déterminer la quantité d'action que les hommes peuvent fournir par leur travail journalier, suivant les différentes manières dont ils emploient leurs forces." *Mémoires de l'Institut national des sciences et arts* 2 (1798): 380–428.

Coutrot, Jean. *L'humanisme économique: les leçons de juin 1936*. Paris: Editions du Centre polytechnicien d'études économiques, 1936.

Cresson, Ernest-Guillaume. *Cent jours de siége à la préfecture de police*. Paris: Plon, Nourrit et cie., 1901.

Cuvier, Georges. *Le règne animal distribué d'après son organisation: pour servir de base à l'histoire naturelle des animaux et d'introduction à l'anatomie comparée. L'introduction, les mammifères et les oiseaux*. Vol. 1. Paris: Deterville, 1817.

———. *Leçons d'anatomie comparée: contenant les généralités et les organes du mouvement des animaux vertébrés*. Vol. 1. Paris: Crochard, 1835.

Dacome, Lucia. "Living with the Chair: Private Excreta, Collective Health and Medical Authority in the Eighteenth Century." *History of Science* 39, no. 126 (2001): 467–500.

Dagognet, François. *Des révolutions vertes: histoire et principes de l'agronomie*. Collection Savoir. Paris: Hermann, 1973.

D'Alméras, Henri. *La vie parisienne pendant le siège et sous la commune*. Paris: Albin Michel, 1927.

Dard, Olivier. "Théoriciens et praticiens de l'économie: un changement de paradigme." In *L'année* 1947, edited by Serge Berstein and Pierre Milza, 75–114. Paris: Presses de Sciences Po, 2000.

———. *Le rendez-vous manqué des relèves des années trente*. Edited by Lucien Bély, Claude Gauvard, and Jean-François Sirinelli. Le noeud gordien. Paris: PUF, 2002.

Davenne, Henri-Jean-Baptiste. *Mémoire au Conseil de surveillance sur la revision du régime alimentaire: séance du 17 avril 1851*. Paris: Paul Dupont, 1851.

David, P. "Nous aimerions savoir en vertu de quel mandat." *Journal des débats*, October 4, 1870.

Degerando, Joseph. "Considérations dans les diverses méthodes à suivre dans l'observation des peuples sauvages." *Revue d'Anthropologie* 2, no. 6 (1883): 152–62.

De Grazia, Victoria. *Irresistible Empire: America's Advance through Twentieth-Century Europe*. Cambridge, MA: Harvard University Press, 2005.

Delabost, Merry. *Régime pénitentiaire; hygiène alimentaire; l'alimentation des détenus au point de vue hygiénique et pénitentiaire*. Paris: Imprimerie Chaix, 1885.

Denormandie, *Ernest. Notes et Souvenirs par M. Denormandie, Senateur. Les Journées de Juin 1848 - Le Siège de Paris - La Commune - L'Assemblée Nationale - Quelques Réflextions*. Paris: Société Anonyme des Publications Périodiques, 1896.

D'Espine, Marc. "Notice sur les variations du poids des prisonniers soumis au régime pénitentiaire." *Annales d'hygiène publique* 1, no. 32 (1844): 71–85.

Desrosières, Alain. *The Politics of Large Numbers: A History of Statistical Reasoning*. Translated by Camille Naish. Boston: Harvard University Press, 1998.

Destanne de Bernis, G. "François Perroux (1903–1987)." In *Biographical Dictionary of Dissenting Economists*, edited by Malcolm Charles Sawyer and Philip Arestis, 494–500. Cheltenham: Edward Elgar, 2000.

Dicaprio, Lisa. "Women Workers, State-Sponsored Work, and the Right to Subsistence during the French Revolution." *Journal of Modern History* 71, no. 3 (1999): 519–51.

Dillon-Kavanagh, G. "La ration." *La Patrie en Danger*, 1870, 1.

Dols, J.-L. *Les enquêtes de consommation en France*. Paris: Editions du CNRS, 1952.

Donzelot, Jacques. *The Policing of Families*. Translated by Robert Hurley. New York: Pantheon, 1979.

Dubreuil, Hyacinthe. "Pour la République industrielle: le salaire ration." *L'Information sociale: Action syndicale, organisation du travail, évolution économique*, April 26 1923, 1–2.

———. *Robots or Men? A French Workman's Experience in American Industry*. Translated by Frances Merrill and Mason Merrill. New York: Harper, 1930.

———. *La chevalerie du travail*. Paris: Bernard Grasset, 1941.

Dumas, Jean Baptiste. *Traité de chimie appliquée aux arts*. 8 vols. Vol. 8. Paris: Bechet Jeune, 1846.

———. *Eloge historique de Henri-Victor Regnault*. Edited by Institut de France. Paris: Firmin-Didot, 1881.

Dumas, Jean Baptiste, and Jean Baptiste Boussingault. *Essai de statique chimique des êtres organisés: leçon professée par M. J. Dumas le 20 août 1841 pour la clôture de son cours à l'Ecole de Médecine*. Third ed. Paris: Fortin, Masson & Cie., 1844.

Dumas, Jean Baptiste, and Henri Milne Edwards. "Sur la composition de la cire des abeilles." *Annales de chimie et de physique*, 3rd series, no. 14 (1845): 400–408.

Durkheim, Emile. *De la division du travail social: étude sur l'organisation des sociétés supérieures*. Paris: Felix Alcan, 1893.

———. *Le suicide, étude de sociologie*. Paris: Felix Alcan, 1897.

Dutton, Paul V. *Origins of the French Welfare State: The Struggle for Social Reform in France, 1914–1947*. Edited by Peter Baldwin, Christopher Clark, James B. Collins, Mia Rodriguez-Salgado, and Lyndal Roper. New Studies in European History. Cambridge: Cambridge University Press, 2002.

Duvergier, J. B. *Collection complète des lois, ordonnances, règlemens et avis du Conseil d'Etat*. Vol. 48, Paris: Sirey, 1848.

Elekund, Robert Burton, and Robert F. Hébert. *Secret Origins of Modern Microeconomics: Dupuit and the Engineers*. Chicago: University of Chicago Press, 1999.

Ellis, Jack D. *The Physician-Legislators of France: Medicine and Politics in the Early Third Republic, 1870–1914*. Cambridge: Cambridge University Press, 1990.

Elwitt, Sanford. *The Third Republic Defended: Bourgeois Reform in France, 1880–1914*. Baton Rouge: Louisiana State University Press, 1986.

Engerman, Stanley L. "Reflections on 'the Standard of Living Debate': New Arguments and New Evidence." In *Capitalism in Context: Essays on Economic Development and Cultural Change in Honor of R. M. Hartwell*, edited by R.M. Hartwell, John A. James, and Mark Thomas, 50–80. Chicago: University of Chicago Press, 1994.

"Famines d'autrefois et disettes d'aujourd'hui." *Le Petit Troyen*, October 14, 1941.

Favre, Jules, Jules Ferry, Jules Simon, Eugène Pelletan, Ernest Picard, Garnier-Pagès, and Emmanuel Arago. "Aux Habitans de Paris." *Journal des débats*, December 13, 1870, 1.

Fogel, Robert. *The Escape from Hunger and Premature Death*, 1700–2100. Cambridge: Cambridge University Press, 2004.

Fondation française pour l'étude des problèmes humains. "Les premiers mois de travail de la Fondation française pour l'étude des problèmes humains." *Cahiers de la Fondation française pour l'étude des problèmes humains* 1 (1943): 7–45.

———. "Compte-rendu des activités de la Fondation française pour l'étude des problèmes humains pour l'année 1943." *Cahiers de la Fondation française pour l'étude des problèmes humains* 2 (1944): 11–117.

Forien de Rochesnard, Jean, and Raymond Granier. "Le rationnement en France, 1939–1950." Typewritten manuscript. Colombes, 1980.

———. "Législation concernant le rationnement en France, 1934–1950." Typewritten manuscript. Colombes, 1981.

Foster, John Bellamy. *Marx's Ecology: Materialism and Nature*. New York: Monthly Review Press, 2000.

Fridenson, Patrick, and Jean-Louis Robert. "Les ouvriers dans la France de la Seconde Guerre mondiale, un bilan." *Mouvement social* 158, Syndicalismes sous Vichy (1992): 117–47.

Frobert, Ludovic. "L'économie politique d'Étienne Antonelli (1879–1971)." *Revue économique* 48, no. 6 (1997): 1529–59.

Gasparin, Adrien de. *Cours d'agriculture*. 5 vols. Vol. 3. Paris: Maison rustique, 1848.

Gautier, Armand. *L'alimentation et les régimes chez l'homme sain et chez les malades*. Second ed. Paris: Masson et Cie., 1904.

Gérando, Joseph Marie de. *Le visiteur du pauvre*. Paris: J. Renouard, 1826.

———. *De l'indigence, des causes de l'indigence, des devoirs imposés à la bienfaisance publique*. Vol. 1, De la bienfaisance publique. Paris: Hachette, 1971.

Gillispie, Charles Coulston. *Science and Polity in France: The End of the Old Regime*. Princeton, NJ: Princeton Univesity Press, 1980.

Girardin, Emile de. *L'abolition de la misère par l'élevation des salaires. Lettres à M. Thiers, Rapporteur de la Commission de l'Assistance et de la Prévoyance publique*. Second ed. Paris: Librairie Nouvelle, 1850.

Glickman, Larry. *The Living Wage: American Workers and the Making of Consumer Society*. Ithaca, NY: Cornell University Press, 1997.

Goldstein, Jan E. "Of Marksmanship and Marx: Reflections on the Linguistic Construction of Class in Some Recent Historical Scholarship." *Modern Intellectual History* 2, no. 1 (2005): 87–107.

Gossez, R. *Les ouvriers de Paris: l'organisation, 1848–1851*. Vol. 24, *Bibliothèque de la Révolution de 1848*. La Roche-sur-Yon: Imprimerie Centrale de l'Ouest, 1967.

Gould, Roger V. *Insurgent Identities: Class, Community, and Protest in Paris from 1848 to the Commune*. Chicago: University of Chicago Press, 1995.

Goullé, Albert. "La question des vivres." *La Patrie en Danger*, September 14, 1870, 3.

Gropius, Walter. "Sociological Premises for the Minimum Dwelling of Urban Industrial Populations." In *The Scope of Total Architecture*, 104–35. New York: Harper & Row, 1943.

Grouvelle, Philippe. *Chauffage et ventilation de la Nouvelle Force à Paris*. Paris: H. Fournier, 1845.

Grynberg, Anne. *Les camps de la honte: les internés juifs des camps français*, 1939–1944. Sciences humaines et sociales. Paris: La Découverte/Poche, 1999.

Guesde, Jules. *La loi des salaires et ses conséquences, suivi d'une réponse à la "Réponse à M. Clemenceau."* Paris: Librarie populaire du Parti Socialiste, 1931.

Guesde, Jules, and Paul Lafargue. *Le programme du Parti ouvrier: ses considérants et ses articles*. Fifth ed. Lille: P. Lagrange, 1899.

Guglielmi, Jean-Louis, and Marguerite Perrot. *Salaires et revendications sociales en France, 1944–1952*. Centre d'études économiques: études et mémoires. Paris: Librairie Armand Colin, 1953.

Guha, Ramachandra. *How Much Should People Consume?* Berkeley: University of California Press, 2006.

Guillaume, James. *L'Internationale: documents et souvenirs (1864–1878)*. Vol. 3, P.-V. Stock: 1909.

Halbwachs, Maurice. "Les besoins et les tendances dans l'économie sociale." *Revue philosophique de la France et de l'étranger* 30, no. 59 (1905): 181–89.

———. *L'évolution des besoins dans les classes ouvrières*. Paris: Félix Alcan, 1933.

Hamlin, Christopher. "The 'Necessaries of Life' in British Political Medicine, 1750–1850." *Journal of Consumer Policy* 29, no. 4 (2006): 373–97.

Harvey, David. *The Limits to Capital*. Second ed. New York: Verso, 1999.

Heilbron, Johan. "Pionniers par défaut? Les débuts de la recherche au Centre d'études sociologiques (1946–1960)." *Revue Française de Sociologie* 32, no. 3 (1991): 365–79.

Hellman, John. *The Knight-Monks of Vichy France*. Montreal: McGill-Queens University Press, 1993.

Heyss, Ignatz Johann. *Der nach dem Lebengewicht des Rindes und selbst nach dessen Quadrat-Maaße normalmäßig zu bestimmende Milchertrag, oder das einfachste und sicherste Mittel zur höchsten Veredlung und Vervollkommung der Rindviehzucht und höchst möglichen Futterausnützung*. Innsbruck: Wagner, 1849.

Hobsbawm, E. J. "The British Standard of Living, 1790–1850." *Economic History Review*, New Series 10, no. 1 (1957): 46–68.

———. "The Standard of Living during the Industrial Revolution: A Discussion." *Economic History Review*, New Series 16, no. 1 (1963): 119–34.

Holmes, Frederick L. *Claude Bernard and Animal Chemistry: The Emergence of a Scientist*. Cambridge, MA: Harvard University Press, 1974.

———. *Lavoisier and the Chemistry of Life: An Exploration of Scientific Creativity*. Edited by William Coleman, David C. Lindberg, and Ronald L. Numbers. Wisconsin Publications in the History of Science and Medicine. Madison: University of Wisconsin Press, 1985.

Horne, Janet R. *A Social Laboratory for Modern France: The Musée Social and the Rise of the Welfare State*. Durham, NC: Duke University Press, 2002.

Horowitz, Daniel. *The Morality of Spending: Attitudes toward the Consumer Society in America, 1875–1940*. Edited by Thomas Bender. New Studies in American Intellectual and Cultural History. Baltimore: Johns Hopkins University Press, 1985.

Howard, Michael. *The Franco-Prussian War: The German Invasion of France*, 1870–1871. London: Rupert Hart-Davis, 1962.

Hurel, A. "Du régime alimentaire dans les maisons centrales." *Annales d'hygiène publique* 2, no. 43 (1875): 336–50.

Husson, Armand. *Les consommations de Paris*. Second ed. Paris: Hachette, 1875.

Inglehart, Ronald, and Jacques René Rabier. "Political Realignment in Advanced Industrial Society: From Class-Based Politics to Quality-of-Life Politics." *Government and Opposition* 24, no. 4 (1986): 456–79.

Institut de Science Economique Appliquée. "Les caractères contemporains de la rémunération du travail." *Cahiers Série B. La rénumération du travail et la politique du salaire* 1 (1945): 1–176.

———. "Salaire et consommation." *Cahiers Série B. La rémunération du travail et la politique du salaire B*, no. 3 (1946): 1–129.

Institut National d'Etudes Démographiques. *Désirs des Français en matière d'habitation urbaine: une enquête par sondage*. Travaux et documents, cahier no 3. Paris: Presses Universitaires de France, 1947.

Jackson, Julian. "'Mal embarqué, bien arrivée': The Strange Story of François Perroux." In *Vichy, Resistance, Liberation: New Perspectives on Wartime France*, edited by Hanna Diamond, and Simon Kitson, 155–70. Oxford: Berg, 2005.

Jarausch, Konrad A., and Michael Geyer. "In Pursuit of Happiness: Consumption, Mass Culture and Consumerism." In *Shattered Past: Reconstructing German Histories*, 269–314. Princeton, NJ: Princeton University Press, 2003.

Jardin, André, and André Tudesq. *Restoration and Reaction, 1815–1848. Cambridge History of Modern France*. Cambridge: Cambridge University Press, 1983.

Jardin, L. "Le juste salaire." In *Le juste prix, Le juste salaire*, 9–24. Paris: J. Mersch, 1896.

Jasanoff, Sheila. "Idioms of Co-production." In *States of Knowledge: the Co-production of Science and Social Order*, edited by Sheila Jasanoff, 1–12. New York: Routledge, 2004.

Jaurès, Jean. *Etudes socialistes*. Paris: Ollendorf, 1902.

Jomard, M. *Discours sur la vie et les travaux du Baron de Gérando*. Paris: Schneider et Lagrand, 1843.

Jordanova, Ludmilla. "Interrogating the Concept of Reproduction in the Eighteenth Century." In *Conceiving the New World Order: The Global Politics of Reproduction*, edited by Faye D. Ginsberg and Rayna Rapp, 369–86. Berkeley: University of California Press, 1995.

Jorland, Gérard. *Une société à soigner: hygiène et salubrité publiques en France au XIXe siècle*. Paris: Gallimard, 2010.

Kaier, Christina. "The Russian Constructivist Flapper Dress." *Critical Inquiry* 28, no. 1 (2001): 185–243.

Kalifa, Dominique. "Conclusion." In *La France des années 1830 et l'esprit de réforme*, edited by Patrick Harismendy, 301–6. Rennes: Presses Universitaires de Rennes, 2006.

Kaplan, Steven L. "The Famine Plot Persuasion in Eighteenth-Century France." *Transactions of the American Philosophical Society* 72, no. 3 (1982): 1–79.

———. "Un laboratoire de la doctrine corporatiste sous le régime de Vichy: l'Institut d'études corporatives et sociales." *Mouvement social* 195 (2001): 35–77.

———. *Le retour du bon pain: une histoire contemporaine du pain, de ses techniques et de ses hommes*. Paris: Perrin, 2002.

Kirk, Neville. "Decline and Fall, Resilience and Regeneration: A Review Essay on Social Class." *International Labor and Working Class History* 57 (2000): 88–102.

Kirn, L. "L'alimentation du soldat." *Journal des Sciences Militaires* 9, no. 14 (1884): 262–95; 419.

Komlos, John. "Preface." In Stature, *Living Standards, and Economic Development: Essays in Anthropometric History*, edited by John Komlos. Chicago: University of Chicago Press, 1994.

Koreman, Megan. *The Expectation of Justice: France, 1944–1946*. Durham, NC: Duke University Press, 1999.

Kuisel, Richard. *Capitalism and the State in Modern France*. Cambridge: Cambridge University Press, 1981.

La Berge, Ann F. *Mission and Method: The Early-Nineteenth-Century French Public Health Movement*. Edited by Charles Webster and Charles Rosenberg. Cambridge History of Medicine. Cambridge: Cambridge University Press, 1992.

Lagabrielle, Juliette. "Le rationnement alimentaire des enfants et des adolescents en France et en Europe." *Cahiers de la Fondation française pour l'étude des problèmes humains* 4 (1945): 35–45.

Lamarck, Jean Baptiste. *Philosophie zoologique*. Paris: Dentu, 1809.

Lamber Adam, Juliette. *Mes illusions et nos souffrances pendant le siège de Paris*. Sixth ed. Paris: Alphose Lemerre, 1906.

Landouzy, L. "De l'irrationnel et de l'insuffisant dans l'alimentation des ouvriers et des employés parisiens: de la nécessité d'une éducation alimentaire donnée dans toutes les écoles." *Revue de la Société scientifique d'hygiène alimentaire et de l'alimentation rationnelle de l'homme* 3, no. 2 (1906): 34–41.

Lapierre, Claude. "Les accords de salaires." *Droit social* 14, no. 1 (1951): 15–22.

———. "L'élaboration du budget-type et la fixation du salaire minimum garanti." *Droit social* 14, no. 6 (1951): 380–90.

Laplace, Pierre-Simon. *Exposition du système du monde*. Sixth ed. 2 vols. Vol. 1. Paris: Bachelier, 1836.

Laporte, Emile, Fabien Magnin, and Isidore Finance. *Le positivisme au Congrès ouvrier*. Paris: Paul Ritti, 1877.

Larousse, Pierre. "Subsistance." In *Grand dictionnaire universel du XIXe siècle: français, historique, géographique, mythologique, bibliographique, littéraire, artistique, scientifique, etc. Tome quatorzième, S–TESTA*, 1175. Paris: Larousse, 1875.

Lassalle, Ferdinand. *Zur Arbeiterfrage: Lassalle's Rede bei der am 16. Apr. 1863, in Leipzig abgehaltenen Arbeiterversammlung*. Liepzig: Lassalle'chen Allgemeinen Deutschen Arbeitervereins.

Latour, Bruno. *Reassembling the Social: An Introduction to Actor-Network Theory*. Clarendon Lectures in Management Studies. Oxford: Oxford University Press, 2007.

Latta, Claude. "Lyon 1834: les victimes de la répression et la seconde révolte des canuts." In *Repression et prison politiques en France et en Europe au XIXe siècle*, edited by Philippe Vigiuer, 21–46. Paris: Creaphis, 1990.

Lavoisier, Antoine Laurent. "Altérations qu'éprouve l'air respiré." In *Oeuvres de Lavoisier: Mémoires de chimie et de physique*, edited by Jean Baptiste Dumas, 676–87. Paris: Imprimerie impériale, 1862.

———. "Résultats extraits d'un ouvrage intitulé De la richesse territoriale du royaume de France." In *Oeuvres de Lavoisier: Rapport à l'Académie, notes et rapport divers, économie politique*, edited by Jean Baptiste Dumas, 403–63. Paris: Imprimerie impériale, 1893.

Lavoisier, Antoine Laurent, and Armand Séguin. "Premier mémoire sur la respiration des animaux." In *Oeuvres de Lavoisier: Mémoires de chimie et de physique*, edited by Jean Baptiste Dumas, 688–703. Paris: Imprimerie impériale, 1862.

Lavollée, Charles. *Exposition universelle internationale de 1889 à Paris. Economie sociale—Section I. Rapport de M. Charles Lavollée*. Edited by Alfred Picard. Rapports du Jury international. Paris: Imprimerie nationale, 1891.

Lawes, J. B. "Animal Chemistry: Pig Feeding." *Journal of the Royal Agricultural Society of England* (1853): 459–544.

Lebret, Louis Joseph, Alexandre Dubois, Jean-Marius Gatheron, Edmond Laulhère, René Moreaux, Gustave Thibon, and François Perroux. "Manifeste d'Economie et Humanisme." *Economie et Humanisme*, February–March 1942, 20–22.

"Le Congrès de la fédération des travailleurs du sous-sol: la question des salaires." *L'Information ouvrière et sociale*, August 29, 1918, 1–4.

Le Gall, Philippe. *A History of Econometrics in France*: From Nature to Models. Business & Economics. New York: Routledge, 2007.

Leblanc, Félix. "Recherches sur la composition de l'air confiné (Mémoire lu à l'Académie des Sciences, le 6 juin 1842)." *Annales de chimie et de physique*, 3rd series, no. 5 (1842): 223–71.

Lebovics, Herman. *The Alliance of Iron and Wheat in the Third French Republic, 1860–1914: The Origins of the New Conservatism*. Baton Rouge: Louisiana State University Press, 1988.

Lecanu, Louis René. "Nouvelles recherches sur l'urine humaine." *Mémoires de l'Académie royale de médecine* 8 (1840): 676–96.

"Le ministère de l'intérieur." Le Siècle, 1850, 1.

Leo XIII. 1891. "Rerum Novarum: On the Condition of the Working Classes." http://www.vatican.va/holy_father/leo_xiii/encyclicals/documents/hf_l-xiii_enc_15051891_rerum-novarum_en.html

Le Play, Frédéric. "Statistique." In *Encyclopédie nouvelle, ou dictionnaire philosophique, scientifique, littéraire et industriel*, edited by Pierre Leroux and Jean Reynaud, 275–81. Paris: Gosselin, 1841.

———. *Instruction sur la méthode d'observation dite des monographies de familles, propre à l'ouvrage intitulé "Les ouvriers européens."* Paris: Société d'économie politique, 1862.

———. *L'organisation du travail selon la coutume des ateliers et la loi du Décalogue*. Tours: A. Mame et fils, 1870.

———. *Les ouvriers européens: études sur les travaux, la vie domestique et la condition morale des populations ouvriers de l'Europe*. Second ed. 6 vols. Vol. 1. Paris: A. Marne et fils, 1879.

———. "The Study of Working Class Families." In *Frédéric Le Play on Family, Work, and Social Change*, edited by Catherine Bodard Silver, 171–74. Chicago: University of Chicago Press, 1982.

Lescure, Michel. "Le manque de ressources: les années 1880–1935." In *La stagnation éco-*

nomique, edited by Maurice Lévy-Leboyer and Jean-Claude Casanova, 256–74. Paris: Gallimard, 1991.

Lévy-Bruhl, Raymond. "Méthodes d'enquête auprès des familles et quelques cas d'application étudiés par l'INSEE." In *Les enquêtes de consommation en France*, edited by J.-L. Dols, 101–12. Paris: Editions du CNRS, 1952.

Lindenberg, Daniel. *Les années souterraines (1937–1947)*. Edited by Thierry Paquot. *L'aventure intellectuelle de la France au XXe siècle*. Paris: La Découverte, 1990.

Loubet del Bayle, Jean-Louis. *Les non-conformistes des années 30: une tentative de renouvellement de la pensée politique française*. Paris: Seuil, 2001.

Lynch, Katherine. *Family, Class, and Ideology in Early Industrial France: Social Policy and the Working-Class Family, 1825–1848*. Edited by David L. Featherman and David I. Kertzer. Life Course Studies. Madison: University of Wisconsin Press, 1988.

Lyubomirsky, Sonja. *The How of Happiness: A Scientific Approach to Getting the Life You Want*. New York: Penguin, 2008.

Magendie, François. "Rapport fait à l'Académie des Sciences au nom de la Commission dite de la gélatine." *Comptes rendus hebdomadaires des séances de l'Académie des Sciences* 13, no. 5 (1841): 237–95.

Maison centrale de Force et de Correction. *Analyse des réponses des directeurs une circulaire ministérielle du 10 mars 1834 sur l'effet des régimes de ces maisons*. Paris: Imprimerie Royale, 1836.

Maitron, Jean, ed. *Prache, Gaston, Louis, Eugène*. CD-Rom. *Dictionnaire biographique du Mouvement ouvrier*. Paris: Editions de l'Atelier, 1997.

Manning, Jonathan. "Wages and Purchasing Power." In Capital Cities at War: Paris, London, Berlin, 1914–1919, edited by Jay Winter and Jean-Louis Robert, 255–85. Cambridge: Cambridge University Press, 1997.

Marchal, Jean. "Salaires et économie d'armistice." In *Travail et salaire*, 87–114. Paris: Editions du Cerf, 1944.

Marchand, Olivier, and Claude Thélot. *Deux siècles de travail en France*. Edited by Etudes. Paris: INSEE, 1991.

———. *Le travail en France, 1800–2000*. Essais et Recherches. Paris: Nathan, 1997.

Marcus, Sharon. *Apartment Stories: City and Home in Nineteenth-Century Paris and London*. Berkeley: University of California Press, 1999.

Marcuse, Herbert. *One-Dimensional Man: Studies in the Ideology of Advanced Industrial Society*. Boston: Beacon Press, 1991.

Marglin, Stephen. "Lessons of the Golden Age: An Overview." In *The Golden Age of Capitalism: Reinterpreting the Postwar Experience*, edited by Stephen Marglin and Juliet B. Schor, 1–38. Oxford: Clarendon Press, 1990.

Martinache, A.-M. *"A travail égal, genre de vie égal": les tendances modernes de la rémunération du travail; salaire vital—salaire familial*. Thèse de doctorat. Edited by Faculté de Droit de Paris. Gentilly: Imprimerie nouvelle, 1926.

Marx, Karl. "Critique of the Gotha Program." 1890. http://www.marxists.org/archive/marx/works/1875/gotha/ch01.htm (accessed February 9, 2011).

———. *Capital*. 4 vols. Vol. 1. London: Penguin, 1990.

Mascarello, L. "Une insulte à la classe ouvrière!" *Le Peuple* 31, no. 318 (1950): 1–3.

Maurel, E. *Traité de l'alimentation et de la nutrition à l'état normal et pathologique*. 5 vols. Vol. 2. Paris: O. Doin, Editeur, 1908.

Mazower, Mark. "Healthy Bodies, Sick Bodies." In *Dark Continent: Europe's Twentieth Century*, 76–103. New York: Vintage, 1998.

McCosh, F. W. J. *Boussingault: Chemist and Agriculturalist*. Chemists and Chemistry. Dordecht: D. Reidel, 1984.

McLaren, Warren. 2009. "Stuff: How Much Do You Really Need?" http://www.treehugger.com/files/2009/02/stuff-how-much-do-you-really-need.php (accessed January 7, 2011).

McLeod, Mary. "Urbanism and Utopia: Le Corbusier from Regional Syndicalism to Vichy." Princeton University, 1985.

McPhee, Peter. *The Politics of Rural Life: Political Mobilization in the French Countryside*, 1846–1852. Oxford: Clarendon Press, 1992.

Mead, Christopher. "Gilbert, Emile(-Narcisse-Jacques)." *Grove Dictionary of Art Online*. http://www.groveart.com (accessed March 30, 2004).

Menchérini, Robert. "Conséquences sanitaires des pénuries alimentaires dans les Bouches-du-Rhône." In *Le temps de restrictions en France* (1939–1949), edited by Dominique Veillon and Jean-Marie Flonneau, 419–32. Paris: Centre national de recherche scientifique, 1996.

Menger, Carl. *Principles of Economics*. Translated by James Dingwell and Bert F. Hoselitz. Auburn, AL: Ludwig von Mises Institute, 2007.

Miller, Peter, and Nikolas Rose. "Governing Economic Life." *Economy and Society* 19, no. 1 (1990): 1–31.

Ministère de l'Agriculture et du Ravitaillement, Bureau Central des Recherches. *Cinquième rapport sur les travaux du Bureau Central des Recherches*. Paris: Imprimerie nationale, 1942.

Ministère de l'Intérieur. *Projet de loi sur les prisons: observations de MM. le Préfets sur ce projet de loi dans des rapports avec l'administration et la police des prisons*. Paris: Imprimerie Royale, 1846.

Mitchell, Timothy. *Carbon Democracy: Political Power in the Age of Oil*. New York: Verso, 2013.

Molinari, Gustave de. *Journal des débats*, October 11 and October 21, 1870.

———. "L'alimentation de Paris pendant le siège," *Revue des Deux-Mondes*, 41st year - second period, no. 91 (1871): 121.

Monfalcon, Jean Baptiste, and Augustin-Pierre Isidore de Polinière. *Traité de la salubrité dans les grandes villes suivi de l'hygiène de Lyon*. Paris: J.-B. Baillière, 1846.

Monseigneur Turinaz. *Le salaire familial et le Cardinal Zigliara*. Nancy: E. Le Chevallier, 1895.

Morgan, Mary S. "Experimental Farming and Ricardo's Political Arithmetic of Distribution." *Working Papers on the Nature of Evidence: How Well Do 'Facts' Travel?* 3, no. 5 (2005): 48.

Morillon, Adolphe. *L'approvisionnement de Paris en temps de guerre: souvenirs et prévisions*. Paris: Perrin, 1888.

Moscati, Ivan. "Were Jevons, Menger, and Walras Really Cardinalists? On the Notion of Measurement in Utility Theory, Psychology, Mathematics, and Other Disciplines, 1870–1910." *History of Political Economy* 45, no. 3 (2013): 373–414.

Moureau, Jules. *Le salaire et des associations coopératives: Etude économique suivie d'une description du Familistère de Guise (Aisne)*. Paris: Guillaumin, 1866.

Mumford, Eric. *The CIAM Discourse on Urbanism, 1928–1960*. Cambridge, MA: MIT Press, 2000.

Munier, Etienne. *Essai d'une méthode générale propre à étendre les connaissances des voyageurs ou recueil d'observations relatives à l'histoire, à la répartition des impôts, au commerce, aux sciences, aux arts et à la culture des terres*. Vol. 2. Paris: Moutard, 1779.

Murphy, Michelle. "Distributed Reproduction." In *Corpus: An Interdisciplinary Reader on Bodies and Knowledge*, edited by Monica J. Casper and Paisley Currah, 21–38. New York: Palgrave Macmillan, 2011.

Mysyrowicz, Ladislas. "Karl Marx, la Première Internationale et la statistique." *Mouvement social* 69 (1969): 51–84.

O'Hara-May, Jane. "Measuring Man's Needs." *Journal of the History of Biology* 4, no. 2 (1971): 249–74.

Oualid, William, and Charles Picquenard. *La guerre et le travail: salaires et tarifs, conventions collectives et grèves; la politique du Ministère de l'Armement et du Ministère du Travail*. Edited by James Shotwell and Carnegie Foundation. Histoire économique et sociale de la Guerre Mondiale. Paris: Presses Universitaires de France, 1928.

Outram, Dorinda. *Georges Cuvier: Vocation, Science, and Authority in Post-Revolutionary France*. Manchester: Manchester University Press, 1984.

———. "Uncertain Legislator: Georges Cuvier's Laws of Nature in Their Intellectual Context." *Journal of the History of Biology* 19, no. 3 (1986): 323–68.

Outward Bound. "Dan's Environmental Corner—Simplicity, What do you really need?" 2009. http://www.outwardbound.co.nz/my-watchmates/newsletter/my-watchmates-march-2009/dans-environmental-corner-simplicity-what-do-you-really-need/ (accessed January 11, 2011).

Pascault, L. "Critique des méthodes employées pour établir la norme alimentaire." *Revue de la Société scientifique d'hygiène alimentaire et de l'alimentation rationnelle de l'homme* 1, no. 3 (1904): 200–7.

Payen, Anselme. "De l'alimentation publique: la viande de boucherie et les conserves alimentaires." *Revue des Deux-Mondes* (1855): 895–914.

———. *Précis théorique et pratique des substances alimentaires et des moyens de les améliorer, de les conserver et d'en reconnaître les altérations*. Fourth ed. Paris: Hachette, 1865.

Péclet, Eugène. *Instruction sur l'assainissement des écoles primaires et des salles d'asile*. Paris: Hachette, 1846.

———. *Nouveaux documents relatifs au chauffage et à la ventilation des établissements publics suivis de nouvelles recherches sur le refroidissement et la transmission de la chaleur, pour servir de supplément à la seconde édition du Traité de la chaleur*. Paris: Hachette, 1853.

———. *Traité de la chaleur considérée dans ses applications aux arts*. Second ed. Paris: Masson, 1861. Third ed., Paris: Masson, 1878.

Pedersen, Susan. *Family, Dependence, and the Origins of the Welfare State: Britain and France, 1914–1945*. Cambridge: Cambridge University Press, 1993.

Pelletier, Denis. *Economie et Humanisme: de l'utopie communautaire au combat pour le tiers monde (1941–1966)*. Histoire. Paris: Cerf, 1996.

Perin, Charles. "Le juste salaire d'après l'encyclique *Rerum novarum*." In *Premiers principes d'économie politique, suivie d'une étude sur le juste salaire d'après l'encyclique Rerum Novarum*, 379–419. Paris: Librairie Victor Lecoffre, 1896.

Perrier, L. "Rapport sur l'alimentation dans l'armée." *Revue de la Société scientifique d'hygiène alimentaire et de l'alimentation rationnelle de l'homme* 3, no. 3 (1906): 469–93.

Perrine, Onésime. *La vie saine et à bon marché*. Fascicule no. 1, *L'alimentation*. Paris: J. Allemane, 1891.

Perrot, Jean-Claude. *Une histoire intellectuelle de l'économie politique, XVIIe–XVIIIe siècle*. Civilisations et sociétés 85. Paris: Ecole des Hautes Etudes en Sciences Sociales, 1992.

Perrot, Michelle. *Les ouvriers en grève: France, 1871–1890*. Edited by Ecole des Hautes Etudes en Sciences Sociales. 2 vols. Vol. 1, Civilisations et sociétés. Paris: Mouton, 1974.

———. "Note sur le positivisme ouvrier." *Romantisme* 8, nos. 21–22 (1978): 201–4.

———. "Les ouvriers, l'habitation et la ville au XIXe siècle." In *La question du logement et le mouvement ouvrier français*, edited by Jean-Paul Flamand, 18–40. Paris: Editions de la Villette, 1981.

Perroux, François. "Introduction." In *Joseph Schumpeter, théorie de l'évolution économique:* recherches sur le profit, le crédit, l'intérêt et le cycle de la conjoncture, 1–148. Paris: Dalloz, 1935.

———. *Capitalisme et communauté de travail*. Paris: Sirey, 1938.

———. *Science de l'homme et science économique*. Fondation française pour l'étude des problèmes humains, Essais sur la science de l'homme. Paris: Librairie de Médicis, 1944.

———. *Economie organisée et économie socialisée*. Groupes Travail. Paris: Domat-Montchrestien, 1945.

———. *Pour une politique des prix*. Groupes Travail. Paris: Domat-Monchrestien, 1945.

Perroux, François, and Yves Urvoy. Renaître, essais. 6 vols. Paris: Librairie de Médicis, 1943.

Persoz. "Expériences sur l'engrais des oies." *Annales de chimie et de physique*, 3rd series, no. 14 (1845): 408–19.

Peschanski, Denis. Vichy 1940–1944: *Contrôle et exclusion*. Edited by Serge Berstein, and Pierre Milza. Questions au XXe siècle. Paris: Editions Complexe, 1997.

Picard, Jean-François. 2001. "The Institut national d'hygiène and Public Health in France, 1940–1946." http://picardp1.ivry.cnrs.fr/INH.html.

Picard, Roger. *Le minimum légal de salaire: les cahiers du socialiste*. Paris: La Librairie du Parti Socialiste, 1913.

———. *Le salaire et ses complements: allocations familiales; assurances sociales*. Edited by Marcel Rivière. Bibliothèque des Sciences Politiques & Sociales. Paris: Librairie des Sciences Politiques et Sociales, 1928.

Pickstone, John. "How Might We Map the Cultural Fields of Science? Politics and Organisms in Restoration France." *History of Science* 37, no. 117 (1999): 347–64.

Picot, Georges. *Les moyens d'améliorer la condition de l'ouvrier. conférence faite à Amiens sous les auspices de la Société industrielle, le 3 décembre 1890*. Paris: Secrétariat de la Société d'Economie Sociale, 1891.

Pietra Santa, Prosper de. Mazas: *études sur l'emprisonnement cellulaire*. Paris: Victor Masson, 1853.

Pilbeam, Pamela. *The 1830 Revolution in France*. London: Macmillan, 1991.

Pinkney, David H. *Decisive Years in France, 1840–1847*. Princeton, NJ: Princeton Unversity Press, 1986.

Pinte, Jean. *Les allocations familiales: Origines. Régime légal*. Paris: Recueil Sirey, 1935.

Pollard, Miranda. *Reign of Virtue: Mobilizing Gender in Vichy France*. Edited by Catharine R. Stimpson. Women in Culture and Society. Chicago: University of Chicago Press, 1998.

Pope, Clayne L. "The Changing View of the Standard-of-Living Question in the United States." American Economic Review 83, no. 2 (1993): 331–36.

Porter, Theodore M. "Objectivity as Standardization: The Rhetoric of Impersonality in Measurement, Statistics, and Cost-Benefit Analysis." In *Rethinking Objectivity*, edited by Alan Megill, 197–237. Durham, NC: Duke University Press, 1994.

———. "Reforming Vision: the Engineer Le Play Learns to Observe Society Sagely." In *Histories of Scientific Observation*, 281–302. Chicago: University of Chicago Press, 2011.

Postone, Moishe. *Time, Labor, and Social Domination: A Reinterpretation of Marx's Critical Theory*. Cambridge: Cambridge University Press, 1996.

Poznanski, Renée. *Les Juifs en France pendant la Seconde Guerre mondiale*. Paris: Hachette, 1997.

Préfecture de la Seine. "Ordonnance concernant la salubrité des habitations." *Ordonnances et arrêtés du Préfet de Police* 67 (1848): 605–10.

Prêté, Henri. "L'action syndicale et la hausse des salaires." *L'Information ouvrière et sociale*, 1918, 1.

Procacci, Giovanna. *Gouverner la misère: La question sociale en France, 1789–1848*. L'Univers historique. Paris: Seuil, 1993.

Proctor, Robert. *Racial Hygiene: Medicine under the Nazis*. Cambridge, MA: Harvard University Press, 1988.

Proudhon, Pierre-Joseph. "Qu'est-ce que la Propriété?" In *Oeuvres Complètes*. Paris: Riviere, 1926.

Quinlan, Sean. "Physical and Moral Regeneration after the Terror: Medical Culture, Sensibility and Family Politics in France, 1794–1804." *Social History* 29, no. 2 (2004): 139–64.

Rabinbach, Anson. *The Human Motor: Energy, Fatigue, and the Origins of Modernity. Berkeley*: University of California Press, 1990.

Rabinow, Paul. *French Modern: Norms and Forms of the Social Environment*. Chicago: University of Chicago Press, 1995.

Radtke, Arne. "La politique salariale de Vichy." In *Les ouvriers en France pendant la Seconde Guerre mondiale*, edited by Denis Peschanski and Jean-Louis Robert, 265–76. Paris: CNRS, 1992.

Randoin, Lucie. *L'alimentation et la vie: les problèmes actuels de l'alimentation*. Les Conférences du Palais de la Découverte. Paris: Université de Paris, 1940.

———. "Les équilibres alimentaires: les aliments de remplacement." *Bulletin de la Société scientifique d'hygiène alimentaire* 29, nos. 1, 2 (1941): 117–53.

Randoin, Lucie, Pierre Le Gallic, Georges Duchêne, Alfred Marc, and Yvonne Perraud. "Méthodes et résultats des enquêtes sur la consommation alimentaire française d'avant-guerre (1937–1938–1939)." In *Les enquêtes de consommation en France*, edited by J.-L. Dols, 19–57. Paris: Editions du CNRS, 1952.

Raynaud, Barthélemy. *Vers le salaire minimum: Etude d'économie et de législation industrielles*. Bibliothèque d'économie politique et de sociologie. Paris: Recueil Sirey, 1913.

Reddy, William. *The Rise of Market Culture: The Textile Trade and French Society*, 1750–1900. Cambridge: Cambridge University Press, 1984.

Reggiani, Andrés Horacio. "Alexis Carrel, the Unknown: Eugenics and Population Research under Vichy." *French Historical Studies* 25, no. 2 (2002): 331–56.

———. *God's Eugenicist: Alexis Carrel and the Sociobiology of Decline*. Edited by Irwin Wall. Monographs in French Studies. New York: Berghahn Books, 2006.

Remusat, Charles de. *Mémoires de ma vie*. Vol. 3. Paris: Plon, 1958.

Renouvier, Charles. *Manuel républicain de l'homme et du citoyen*. Second ed. Paris: Garnier frères, 1981.

Reynaud, Paul. *Une arme de guerre, le rationnement, par Paul Reynaud, ministre des Finances, allocution radiodiffusée le 29 février 1940*. N.p.: n.p., n.d.

Rials, Stéphane. *Nouvelle histoire de Paris: De Trochu à Thiers, 1870-1873*. Edited by Association pour la publication d'une histoire de Paris. *Nouvelle histoire de Paris*. Paris: Hachette, 1985.

Ricardo, David. On the Principles of Political Economy and Taxation. Edited by Piero Sraffa, and Maurice Dobb. 11 vols. Vol. 1, *The Works and Correspondance of David Ricardo*. Cambridge: Cambridge University Press, 1951.

Richet, Charles. "Études sur le rationnement alimentaire à l'Académie de Médecine." *Bulletin de la Société scientifique d'hygiène alimentaire* 29, nos. 1, 2 (1941): 72–79; 109.

Rioux, Jean-Pierre. *La France de la Quatrième République: l'ardeur et la nécessité*. 2 vols. Vol. 1, Nouvelle histoire de la France contemporaine. Paris: Seuil, 1980.

Robbins, Lionel. *An Essay on the Nature and Significance of Economic Science*. Second ed. London: Macmillan, 1945.

Roberts, Lissa. "Condillac, Lavoisier, and the Instrumentalization of Science." *The Eighteenth Century: Theory and Interpretation* 33, no. 3 (1992): 252–71.

Rocke, Alan J. "Organic Analysis in Comparative Perspective: Liebig, Dumas, and Berzelius, 1811–1837." In *Instruments and Experimentation in the History of Chemistry*, edited by Frederick L. Holmes and Trevor H. Levere, 273–310. Cambridge, MA: MIT Press, 2000.

———. *Nationalizing Science: Adolphe Wurtz and the Battle for French Chemistry*. Edited by Jed Buchwald. Transformations: Studies in the History of Science and Technology. Cambridge, MA: MIT Press, 2001.

Romanet, Emile. *Le salaire familial*. Grenoble: Imprimerie Aubert, 1918.

Rosanvallon, Pierre. *Le moment Guizot*. Paris: Gallimard, 1985.

Rose, Nikolas. *The Politics of Life Itself: Biomedicine, Power, and Subjectivity in the Twenty-First Century*. Princeton, NJ: Princeton University Press, 2007.

Rosenbloom, Stephanie. "But Will It Make You Happy?" *New York Times*, August 8, 2010.

Rössner, Stephen. "The Wine Merchant as Weight Watcher." *Obesity Research* 4, no. 6 (1996): 577–78.

Rougerie, Jacques. "Remarques sur l'histoire des salaires à Paris au XIXe siècle." *Le Mouvement social* 63 (1968): 71–108.

Roux, Jean-Charles. *Le régime alimentaire des écoliers au point de vue pratique*. Conférences du Musée Pédagogique. Paris: Imprimerie nationale, 1909.

Rudischhauser, Sabine. "Salaire minimum et libre concurrence: le rôle de la demande publique dans la constitution d'un marché national du travail France-Allemagne 1890–1914." In *Le travail et la nation: histoire croisée de la France et de l'Allemagne*, edited by Bénédicte Zimmerman, Claude Didry, and Peter Wagner, 199–221. Paris: Editions de la Maison des sciences de l'homme, 1999.

Salais, Robert, and Luciani, Jean. "Matériaux pour la naissance d'une institution: l'Office du travail (1890–1900)." *Genèses* 2, no. 2 (1999): 83–108.

Savoye, Antoine. "La monographie sociologique: jalons pour son histoire (1855–1974)." Les *Etudes Sociales* 131–32 (2000): 11–46.

Schabas, Margaret. *The Natural Origins of Economics*. Chicago: University of Chicago Press, 2005.

Schacherer. "Budget des travailleurs." *La Ruche Populaire* (1841): 1–10.

Schiebinger, Londa L. *Nature's Body: Gender in the Making of Modern Science*. New Brunswick, NJ: Rutgers University Press, 2004.

Schindler, Charles. L'*alimentation du soldat en campagne: la ration de guerre et la préparation rapide des répas en campagne*. Edited by Henri Charles-Lavauzelle. Petite bibliothèque de l'armée française. Paris: Imprimerie Librarie Militaire, 1887.

Schmitt, Hermann. *Die Steuerfreiheit des Existenzminimums: ein Beitrag zur Theorie der Einkommensteuern*. Leipzig: Dunker & Humblodt, 1877.

Schneider, William H. *Quality and Quantity: The Quest for Biological Regeneration in Twentieth-Century France*. Edited by Charles Webster and Charles Rosenberg. Cambridge History of Medicine. Cambridge: Cambridge University Press, 1990.

Schueller, Eugène. *Du salaire-subsistance au salaire production: Le salaire proportionnel; communication présentée à la séance du 6 novembre 1942 du Comité National de l'Organisation Française tenue en commun avec la Société des Ingénieurs civils de France*. Paris: Imprimerie union, 1942.

Scott, Joan Wallach. "'L'ouvrière! Mot impie, sordide': Women Workers in the Discourse of French Political Economy, 1840–1860." In *Gender and the Politics of History*. New York: Columbia University Press, 1999.

Seccombe, Wally. Weathering the Storm: *Working-Class Families from the Industrial Revolution to the Fertility Decline*. New York: Verso, 1993.

Segard, Dr. "Hygiène et solidarité." *Les Temps nouveaux* 10, no. 10 (1904): 1–2.

Sen, Amartya. *Inequality Reexamined*. Oxford: Oxford University Press, 1995.

Séquillon, Y. *L'évolution des salaires*. Paris: Imprimerie A. Humblot, 1950.

Serman, William. "French Mobilization in 1870." In *On the Road to Total War: The American Civil War and the German Wars of Unification, 1861–1871*, edited by Stig Förster, and Jörg Nagler, 283–94. Cambridge: Cambridge University Press, 1997.

Sewell, William H., Jr. *Work and Revolution in France: The Language of Labor from the Old Regime to 1848*. Cambridge: Cambridge University Press, 1980.

Shapiro, Ann-Louise. *Housing the Poor of Paris, 1850–1902*. Madison: University of Wisconsin Press, 1985.

Shovlin, John. "The Cultural Politics of Luxury in Eighteenth-Century France." *French Historical Studies* 23, no. 4 (2000): 577–606.

Simmons, Dana. "Waste Not, Want Not: Excrement and Economy in Nineteenth-Century France." *Representations* 96 (2006): 73–98.

Sismondi, Jean Charles Léonard de. *Nouveaux principes d'économie politique, ou De la richesse dans ses rapports avec la population*. Vol. 1. Paris: Delaunay, 1819.

Sixième Congrès national des syndicats de France: compte rendu des travaux du congrès tenu à Nantes du 17 au 22 septembre 1894. Nantes: Imprimerie George Schwob et fils, 1894.

Société d'Etudes Economiques et Documentaires. "Aperçus véridiques sur la farce du minimum interprofessionnel." *Bulletin economique* (1950).

Soper, Kate. *What Is Nature?* Oxford: Blackwell, 1995.

———. "The Politics of Nature: Reflections on Hedonism, Progress, and Ecology." *RCNS* 10, no. 2 (1999): 47–70.

———. "Conceptualizing Needs in the Context of Consumer Politics." *Journal of Consumer Policy* 29, no. 4 (2006): 355–72.

Spang, Rebecca. "'And they Ate the Zoo': Relating Gastronomic Exoticism in the Siege of Paris." *MLN* 107, no. 4, French Issue (1992): 752–73.

Spary, E. C. *Feeding France: New Sciences of Food*, 1760–1815. Cambridge Social and Cultural Histories. Cambridge: Cambridge University Press, 2014.

Statistique générale de la France. *Salaires et coût de l'existence à diverses époques, jusqu'en 1910*. Paris: Imprimerie nationale, 1911.

Steffen, W., J. Grinevald, P. Crutzen, and J. McNeill. "The Anthropocene: Conceptual and Historical Perspectives." *Philos Transact A Math Phys Eng Sci* 369, no. 1938 (2011): 842–67.

Stigler, George J., and Gary S. Becker. "De Gustibus Non Est Disputandum." *American Economic Review* 67, no. 2 (1977): 76–90.

Stocking, George. "French Anthropology in 1800." *Isis* 55, no. 2 (1964): 134–50.

Stovall, Tyler. "From Red Belt to Black Belt: Race, Class and Urban Marginality in Twentieth-Century Paris." In *The Color of Liberty: Histories of Race in France*, edited by Susan Peabody, and Tyler Stovall, 351–70. Durham, NC: Duke University Press, 2003.

"Subsistances du peuple." *L'Atelier* 6 (1846): 202.

Sweets, John F. *Choices in Vichy France: The French under Nazi Occupation*. Oxford: Oxford University Press, 1986.

Taithe, Bertrand. *Defeated Flesh: Medicine, Welfare and Warfare in the Making of Modern France*. Lanham, MD: Rowman & Littlefield, 1999.

———. *Citizenship and Wars: France in Turmoil*, 1870–1871. New York: Routledge, 2001.

Tang, Jeffrey. "The Clean-Slate Guide to Simplicity." 2010. http://zenhabits.net/clean-slate/ (accessed January 11, 2011).

Tapié, Jean, and Menier Tapié. "Les mesures de salut public." Paris, 1870.

Tardieu, Ambroise. "Pénitentiaire (système)." In *Dictionnaire d'hygiène publique et de salubrité*, 241–91. Paris: J.-B. Ballière, 1862.

———. "Subsistances." In *Dictionnaire d'hygiène publique et de salubrité*, 202–15. Paris: J.-B. Ballière, 1862.

Teige, Karel. *The Minimum Dwelling.* Translated by Eric Dluhosch. Cambridge, MA: MIT Press, 2002.

Terjanian, Anoush. "Doux Commerce and Its Discontents: Slavery, Piracy, and Monopoly in Eighteenth-Century France." Johns Hopkins University, 2005.

Thompson, Victoria E. *The Virtuous Marketplace: Women and Men, Money and Politics in Paris,* 1830–1870. Johns Hopkins University Studies in Historical and Political Science. Baltimore: Johns Hopkins University Press, 2000.

Tiersten, Lisa. *Marianne in the Market: Envisioning Consumer Society in Fin-de-Siècle France.* Berkeley: University of California Press, 2001.

Tombs, Robert. "The Wars against Paris." In *On the Road to Total War: The American Civil War and the German Wars of Unification,* 1861–1871, edited by Stig Förster and Jörg Nagler, 541–64. Cambridge: Cambridge University Press, 1997.

Treitel, Corinna. "Max Rubner and the Biopolitics of Rational Nutrition." Central European *History* 41, no. 1 (2008): 1–25.

Trémolières, Jean, and F. Vinit. "Enquête sur l'état de nutrition de la population, Paris (hiver 1950–1951)." *Bulletin de l'Institut National d'Hygiène* 7, no. 3 (1952): 613–43.

Trémolières, Jean, F. Vinit, and Y. Serville. "Etude sur la ration alimentaire type à préconiser pour le Français." *Bulletin de l'Institut National d'Hygiène* 7, no. 4 (1952): 767–825.

Tribe, Keith. *Strategies of Economic Order: German Economic Discourse,* 1750–1950. Edited by Quentin Skinner, Lorraine Daston, Wolf Lepenies, Richard Rorty, and J. B. Schneewind. Ideas in Context. Cambridge: Cambridge University Press, 1995.

———. "Continental Political Economy: From the Physiocrats to the Marginal Revolution." In *The Cambridge History of Science: Vol. 7, The Modern Social Sciences,* edited by Theodore M. Porter and Dorothy Ross, 154–70. Cambridge: Cambridge University Press, 2003.

Triebel, Armin. "Moral und Ökonomie: Zur modernen Semantik des Lebensstandards." In *Europäische Konsumgeschichte: Zur Gesellschafts- und Kulturgeschichte des Konsums (18. bis 20. Jahrhundert),* edited by Hannes Siegrist, Hartmut Kaelble, and Jurgen Kocka, 365–92. Frankfurt: Campus Verlag, 1997.

"Troubles à Paris pour le pain." *L'Atelier* 7, no. 1 (1846): 394.

"Une conquête ouvrière: le salaire minimum vital." *Inter-France,* December 8, 1941.

"Une question d'actualité: le minimum d'existence." *Les Nouveaux Temps,* March 12, 1941.

Union Nationale des Caisses d'Allocations Familiales. *Les causes matérielles de la dénatalité. La Famille d'Aujourd'hui. Sa formation. Son développement. Enquête de l'Union Nationale des Caisses d'Allocations Familiales.* Paris: FNCSS/UNCAF, 1948.

Vatin, François. *Le travail, économie et physique,* 1780–1830. Edited by Françoise Balibar, Jean-Pierre Lefebvre, Pierre Machery, and Yves Vargas. Philosophies. Paris: Presses Universitaires de France, 1993.

Veillon, Dominique. *Vivre et survivre en France,* 1939–1947. Histoire Payot. Paris: Editions Payot & Rivages, 1995.

Veillon, Dominique, and Jean-Marie Flonneau. "Introduction." In *Le temps des restrictions en France* (1939–1949), edited by Dominique Veillon and Jean-Marie Flonneau. Paris: Centre national de la recherche scientifique, 1996.

Vernois, Maxime. *Etat hygiénique des lycées de l'empire en 1867, extrait du rapport présenté à S. Exc. le Ministre de l'Instruction Publique.* Paris: J.-B. Baillière et fils, 1868.

Vernon, James. *Hunger: A Modern History*. Cambridge, MA: Harvard University Press, 2007.

Vigier, Philippe. 1848, *les français et la république: la vie quotidienne*. Paris: Hachette, 1998.

Villermé, Louis-René. *Tableau de l'état physique et moral des ouvriers employés dans les manufactures de coton, de laine et de soie*. Vol. 2. Paris: Jules Renouard, 1840.

Villey, Edmond. *La question des salaires ou la question sociale*. Paris: Larose et Forcel, 1887.

"Vital, ale." In *Dictionnaire de L'Académie française*, 730. Paris: Hachette, 1932.

"Vite le salaire minimum!" *Paris Soir*, 1941.

Voldman, Danièle. "Le logement: crise, pénurie, ou restrictions?" In *Le temps des restrictions en France* (1939–1949), edited by Dominique Veillon and Jean-Marie Flonneau, 377–88. Paris: Centre national de recherche scientifique, 1996.

Wagner, Peter. " 'An Entirely New Object of Consciousness, of Volition, of Thought': The Coming into Being and (Almost) Passing Away of 'Society' as a Scientific Object." In *Biographies of Scientific Objects*, edited by Lorraine Daston, 132–57. Chicago: University of Chicago Press, 2000.

Webb, Sidney, and Beatrice Potter Webb. *A History of Trade Unionism*. London: Longmans Green, 1896.

Weill, Joseph. *Contribution à l'histoire des camps d'internement dans l'Anti-France*. Etudes et Monographies. Paris: Editions du Centre de Documentation Juive Contemporaine, 1946.

Weiner, Amir, ed. "Landscaping the Human Garden: Twentieth-Century Population Management in a Comparative Framework," Stanford, CA: Stanford University Press, 2003.

Weiss, John Hubbel. *The Making of Technological Man: The Social Origins of French Engineering Education*. Cambridge, MA: MIT Press, 1982.

Weisz, George. *The Medical Mandarins*. Oxford: Oxford University Press, 1995.

Wicksell, Knut. *Lectures in Political Economy*. Translated by E. Classen. Vol. 1, General Theory. Auburn, AL: Ludwig von Mises Institute, 2007.

Williams, Elizabeth A. *The Physical and the Moral: Anthropology, Physiology, and Philosophical Medicine in France*, 1750–1850. Cambridge History of Medicine. Cambridge: Cambridge University Press, 1994.

Williams, Raymond. "Ideas of Nature." In *Culture and Materialism: Selected Essays*, 67–85. London: Verso, 2005.

Williams, Rosalind. *Dream Worlds: Mass Consumption in Late Nineteenth-Century France*. Berkeley: University of California Press, 1981.

Wise, M. Norton. "Mediations: Enlightenment Balancing Acts, or the Technologies of Rationalism." In *World Changes: Thomas Kuhn and the Nature of Science*, edited by Paul Horwich, 207–58. Cambridge, MA: MIT Press, 1993.

Wise, Norton. "Work and Waste: Political Economy and Natural Philosophy in Nineteenth-Century Britain." *History of Science* 27, no. 77 (1989): 263–301.

Yeo, Eileen Janes. "Social Surveys in the Eighteenth and Nineteenth Centuries." In *Genres and Objects of Social Inquiry*, edited by Theodore M. Porter and Dorothy Ross, 83–99. Cambridge: Cambridge University Press, 2003.

Zabiecka, Bozena. "Expériences de salaires minima à l'étranger et leurs repercussions économiques et sociales." In *La Rémunération du travail et la politique du salaire*, 67. Paris: I.S.E.A., 1945.

———. "Les salaires minima en France." In *Cahiers série B: la rémunération du travail et la politique du salaire*, edited by Institut de Science Economique Appliquée, 56. Paris: I.S.E.A., 1945.

Zola, Emile. *Germinal*. Edited by A. Lanoux and H. Mitterand. 3 vols. Vol. 3. Paris: Gallimard, 1964.

INDEX